CYBERSECURITY

SAFEGUARDING OUR DIGITAL WORLD

JUSTIN K. KOJOK

JUSTIN K. KOJOK

Printed Worldwide
First Printing 2024
First Edition 2024

ISBN: 978-1-966348-00-9 (Hardcover)
ISBN: 978-1-966348-01-6 (Paperback)
ISBN: 978-1-966348-02-3 (eBook)

10 9 8 7 6 5 4 3 2 1

Contact: author@kojokbooks.com
Website: https://kojokbooks.com

Disclaimer Notice:
Please note the information contained within this book is for educational purposes only. All effort has been made to present accurate, up to date, and reliable, complete information.
The content within this book has been derived from various sources.
By reading this book, the reader agrees that under no circumstances is the author responsible for any losses, direct or indirect, which are incurred as a result of the use of information contained within book, including, but not limited to, errors, omissions or inaccuracies.

CYBERSECURITY

SAFEGUARDING OUR DIGITAL WORLD

*To **Mr. Jacob Namso**,*

My Middle School Headteacher, my North Star.

You saw the glimmer of possibility in a quiet student's eyes when even I was blind to my own light. I was young and lived in a world of uncertainty, but your steady hand pointed toward horizons I hadn't dared to imagine.

Your words were seeds planted in fertile soil. Today, those seeds have grown into thriving forests, and every branch reaches skyward because you taught me to look up.

My work in the literary world proves that sometimes all a young mind needs is one person who refuses to accept the boundaries they've drawn for themselves.

Thank you for showing me that potential is not about where we are but who we dare to become.

Your faith lit a path I now walk confidently, and I am forever grateful for that.

BOOKS BY JUSTIN K. KOJOK

1. *DRIVING FOR JUSTICE*
2. *INTERCESSOR: A Collection Of My Fervent Prayers*
3. *WHISPERS IN THE VINEYARD*

PLEASE SCAN THIS QR CODE TO BUY THESE BOOKS.

CONTENTS

PREFACE

Cybersecurity is the practice of safeguarding devices, networks, software, and data from digital threats. In this era, digital threats are not minor issues anymore. It's now a priority for companies, governments, and individuals. Digital threats like ransomware and state-sponsored attacks are on the rise every day. It is pertinent to have cybersecurity measures in place to protect infrastructure, intellectual property, and personal information. This book, "Cybersecurity: Safeguarding Our Digital World," intends to offer an examination of today's cybersecurity environment by focusing on technologies and effective defense strategies against continually changing threats.

This book aims to provide readers with an insight into the strategic elements of cybersecurity in a comprehensive manner. Each chapter explores how to establish strong defenses against cyber threats and fosters a sturdy cybersecurity framework. It covers everything from the basics of cybersecurity to the technological progressions. It encompasses a broad spectrum of topics tailored to empower readers with the information required to maneuver through the sophisticated realm of cyber defense.

Challenges of Cybersecurity

The quick development of technology has completely changed our way of life and the way we interact with one another on a personal and professional level. Despite these developments, a plethora of cybersecurity problems has also emerged.

Today, businesses adopt cloud computing technology, enlist 5G networks, and integrate Artificial Intelligence (AI). With this integration, they are simultaneously opening themselves up to vulnerabilities that cyber attackers are keen to capitalize upon. The rising prevalence of Internet of Things (IoT) gadgets, in conjunction with the increasing interconnectedness of networks, is broadening the target area for malevolent entities. It is noted that the rise in cyber threats is posing challenges for organizations globally as they become more frequent and sophisticated.

This book will help readers to understand the progression of dangers and the different strategies employed by perpetrators to breach networks and compromise data while and cause disruptions of business operations. The sections discussing malware, ransomware, and social engineering techniques will comprehensively examine how these attacks have advanced over time and their ramifications on businesses and economies. Understanding the beginnings and development of these dangers will help readers better anticipate and defend against attacks.

The Role of Emerging Technologies
The book also explores how new technologies affect cybersecurity defenses and attacks. It explores the applications of AI and machine

learning as well as developments in Blockchain and quantum computing.

These innovative technologies present a scenario of opportunities and obstacles. AI enables the creation of threat detection mechanisms capable of real-time data analysis for anomaly detection, while cyber criminals exploit AI to streamline attacks and bypass conventional security protocols. This calls for organizations and technology experts to embrace AI driven defense mechanism.

The sections that focus on technologies delve into how these cutting-edge toolsets can enhance cybersecurity measures while also acknowledging the potential threats they bring about. For instance, the part about Blockchain technology emphasizes its ability to safeguard data transactions and validate exchanges. It also delves into the weaknesses that may surface as Blockchain gains acceptance. Moreover, the book delves into quantum computing and its capacity to shake up encryption protocols, which could greatly impact data protection in many ways.

Cyber Resilience

The book also highlights the significance of cyber resilience. Cyber resilience is more than just stopping attacks; it also includes how well an organization can bounce back and adjust after cyber incidents occur. The NIST clearly indicates, "Cyber resilience isn't about getting rid of all risks, but about making sure organizations can keep functioning even if a cyberattack is successful." These sections discuss how to handle incidents and respond to them effectively, as well as the strategies for maintaining business operations during disruptions. It offers guidance on how to recover from these incidents and how to establish a robust business security posture that can endure and bounce back from cyber-attacks.

To enhance cybersecurity effectively, it demands a combined effort from all parties involved. The text explores the importance of businesses coming together through partnerships between the private sectors and engages in information exchange and joint defense strategies. It probes into the involvement of bodies, industry groups, and global organizations in orchestrating countermeasures against cyber risks. By promoting teamwork and open communication among stakeholders, companies can establish a robust defense against threats from nation-state cyber offenders and activism-driven hackers.

Enhance Skills and Learning

This book emphasizes on the critical role of education and workforce training in tackling cybersecurity threats head on. The low supply and high demand for cybersecurity experts equilibrium poses challenges in the area of recruitment and retention for organizations striving to fortify their networks. The (ISC) ² reports indicate that a severe shortage of cybersecurity professionals globally ranks among the concerns for organizations today as millions of positions remain vacant.

This book has sections that discuss cybersecurity education and the development of the workforce and recommendations for preparing cybersecurity experts effectively. They delve into areas such as the importance for certifications and the ongoing process of learning while also tackling the significance of diversity within the workforce in detail. This offers readers an understanding of ways they can contribute to bridging the gap in cybersecurity skills. Moreover, the book delves into how companies can foster a culture that prioritizes security by integrating cybersecurity practices into their core values.

Ethical and Legal Issues

As cyber threats continue to advance in complexity and variety, the legal complications related to cybersecurity are also evolving. *Cybersecurity: Safeguarding Our Digital World* incorporates sections on the accountability of governments and corporations in the face of data breaches. It also delves into the structures overseeing cybersecurity on a global scale. From the General Data Protection Regulation (GDPR) to agreements on cybersecurity, the book offers an outline of the rules that companies should navigate to uphold compliance in today's worldwide marketplace.

A Roadmap for the Future

In the fast changing field of cybersecurity, staying ahead of threats calls for a thorough awareness of the opportunities and difficulties defining our digital environment, not only reactive actions. This book provides a road map to enable readers negotiate this challenging and always shifting terrain. By the time you get to the end, you will have not only acquired knowledge but also a clear direction on how to put it to use successful. Every chapter is designed to address the challenges as well as the opportunities influencing the goal of safeguarding the digital domain. From knowledge of fundamental ideas like the CIA triad to investigation of innovative technologies, including artificial intelligence, Blockchain, and quantum cryptography, the book offers relevant, practical, forward-looking insights. Case studies and practical examples show how both companies and people might create strong defenses and adjust to new challenges. Whether you are a seasoned IT professional, a cybersecurity practitioner, a student starting your path into this important field, or another, this book provides the tools, technologies, and ideas that will define digital defense going forward. It provides viewpoints that motivate proactive thinking and creativity that transcends theory. Following this road map will help readers not only react to cyberattacks but also foresee and reduce them, so preserving our shared digital environment for the next generations.

CHAPTER 1

FOUNDATION OF CYBERSECURITY

Note: all references in this chapter are on page 393 - 394

This chapter will establish the foundation for understanding cybersecurity. Confidentiality, Integrity, and Availability are the fundamental principles that serve as the foundation of digital security. This is also called the CIA triad. Organizations safeguard their data by these principles.

The CIA triad ensures the security of their systems. We will also examine the progression of cybersecurity. We will look at the fundamentals of safeguarding credentials, and multilayered defense strategies. We will explore the increasing complexity of cyber threats. These include malware, phishing, ransomware, and social engineering. You will have a thorough understanding of the fundamental concepts that underpin modern cybersecurity practices. All readers, irrespective of their level of competence would have a clear understanding of the fundamental components of cybersecurity, thereby laying the groundwork for more in-depth discussions on advanced topics. We will examine the role of encryption, multi-factor authentication, and firewalls.

1.1 The Evolution of Cybersecurity

Over the past few years, cybersecurity has developed dramatically. Technological developments and the growing frequency and complexity of cyber threats have been driving this. From the early days of basic security measures like passwords to today's advanced, the cybersecurity domain has evolved into a vital pillar assuring the protection of people, businesses, and even countries.

Early Days: The Birth of Cybersecurity Concepts

Cybersecurity began in the 1960s when the first computer networks, such as Advanced Research Projects Agency Network (ARPANET), were the forerunners of the contemporary internet. In those days, security issues were simple and mostly related to stopping illegal access to systems by simple password protection. But as networks grew and got more linked, so did the necessity for increasingly sophisticated security policies. The Morris Worm in 1988, which attacked roughly 10% of the computers linked to the internet at that time, was among the first and most important incidents influencing the evolution of cybersecurity. The incident highlighted the weaknesses in networked systems and spurred debates on the need to put more strict protection mechanisms into use. Renowned security analyst Peter Neumann considered this and said, "The Morris Worm served as a crucial warning about the vulnerabilities present in networked systems."[1]

The 1990s: Firewalls and Antivirus Programs

The emergence of the World Wide Web in the 1990s brought more problems. Cyberattacks were more likely given the growing volume of data being exchanged across networks. Firewalls were created to monitor and manage network traffic, therefore separating trusted internal networks from untrusted external networks. Firewalls are still absolutely essential parts of modern cybersecurity architecture. Antivirus applications also first emerged during this time; Symantec released the Norton Antivirus program in 1991. This advancement transformed the way people and companies shielded themselves from viruses and trojans[2] as well as from malware.

Notwithstanding these developments, the 1990s were characterized by a growing flood of cyber threats, including Distributed Denial-of-Service (DDoS) attacks, which overwhelmed online services with traffic and thereby crippled them. Especially the Melissa Virus of 1999 showed how increasingly capable thieves are to wreak havoc on world systems. They quickly spread throughout email systems and caused extensive harm and finally resulted in millions of dollars in losses.

The Early 2000s: Encryption and Identity Protection

Early in the 2000s, encryption technologies became increasingly popular, especially with the growth of online transactions and e-commerce. Secure Sockets Layer (SSL) protocols, which encrypt data in transit, became an indispensable weapon for protecting communications between consumers and websites. Leading authority on cryptography Bruce Schneier observed, "encryption became the cornerstone of secure communications, especially as e-commerce and digital financial transactions grew in importance."[3]

Furthermore, identity theft started to be a big issue at this time, which led to the acceptance of two-factor authentication (2FA) and other steps to safeguard personal digital identities. These steps addressed the need for new, strong ways of confirming users outside of basic passwords, which had grown ever more susceptible to hacking. More sophisticated identity verification systems like biometric authentication originated in this age.

2010s: A Strategic Focus on Cybersecurity

In the 2010s, people's conceptions about cybersecurity evolved greatly. Cybersecurity domain evolved from solving technical issues to a strategic status by companies, governments, and individuals.

High-profile cyberattacks demonstrating the severity of cybercrime included the Target data breach in 2013 affecting 41 million consumers and the Equifax hack in 2017 threatening the personal data of 147 million people. This was one of the largest data breaches in U.S. history at the time. These events cost a lot of money and hurt the companies' reputations. This shows that all businesses need to make cybersecurity a top priority.[4]

In addition, cyberattacks paid for by nation states became a major issue, and governments got more involved in both offensive and defensive cyber operations. As an example, the Stuxnet worm, which was speculated to have been made by the US and Israel, attacked Iran's nuclear facilities in 2010. It was one of the first cases of cyber warfare. A computer expert named Scott Shackelford says, "The rise of nation-state actors in cyber warfare has escalated the stakes, pushing cybersecurity to the forefront of national defense strategies."[5]

During this time, artificial intelligence (AI) and machine learning (ML) started to be included into cybersecurity instruments. This made more advanced systems of threat detection and response possible. Managing the rising amount of cyber threats now depends on these technologies. Their predictive powers enable companies to prevent attacks before they start.

The Present Day: Quantum Computing and IoT Risks

As we move into the 2020s, the fast growth of quantum computing brings both new opportunities and problems to cybersecurity that have never been seen before. Many of the encryption techniques now in use are likely to be broken by quantum computers. This has driven scientists to research and develop quantum-resistant encryption algorithms to guard future systems and applications. Experts agree that the cybersecurity scene will change quickly to meet these issues; some even expect a major redesign of our approach to data security.[6]

Apart from quantum computers, the explosion of Internet of Things (IoT) devices has enlarged the attack area for cybercrime. Many times, without strong security measures, IoT devices are easily hacked. The 2016 Mirai botnet assault showed the possible dangers connected with this new generation of technology since it utilized IoT devices to start a large DDoS attack. Securing IoT devices will be among the most important issues facing us as their number increases in the years ahead.

The Need for Proactive Cybersecurity

The development of cybersecurity has been shaped by the ongoing interaction between new threats and better technology. From simple passwords in the early days of computers to systems with AI today, cybersecurity has grown into a complicated field with many layers that affect every part of modern life. As new technologies like quantum computing come out, they will likely change the future of cybersecurity. These technologies will bring problems and solutions to keep the digital world safe. It is important that the strategies used to fight risks change along with them so that people, businesses, and countries can all stay safe in a world that is becoming increasingly linked.

1.2 Key Concepts and Terminology

To get around in the complicated world of digital safety, we need to understand key ideas and terms in cybersecurity. These basic ideas are what cybersecurity practices and strategies are built on. They help experts talk to each other clearly and come up with complete ways to protect themselves from new cyber threats.

Confidentiality, Integrity, and Availability (CIA Triad)

The CIA triad is one of the most foundational models in cybersecurity, representing three essential principles: Confidentiality, Integrity, and Availability. These three concepts provide a comprehensive framework for understanding and addressing the core security needs of any information system.

1. **Confidentiality** is the assurance that sensitive information is only accessible to authorized individuals. Data breaches often violate confidentiality, leading to the unauthorized disclosure of sensitive information. For example, in the 2013 Target data breach, confidential customer information, including credit card numbers, was accessed by cybercriminals. This illustrates the importance of maintaining confidentiality to protect personal and corporate data.[1]

   According to cybersecurity expert Matthew Bishop, "Confidentiality ensures that sensitive information remains inaccessible to unauthorized parties, protecting the privacy of individuals and organizations."[2]

1. **Integrity** refers to the accuracy and reliability of data. Integrity ensures that information is not altered or tampered with by unauthorized individuals during storage or transmission. Maintaining integrity is particularly critical in industries like finance, where even minor data alterations can lead to significant consequences. In 2015, hackers manipulated transaction details in the Bank of Bangladesh cyber heist, showcasing how integrity violations can have substantial financial impacts.[3]

2. **Availability** ensures that information and systems are accessible to authorized users when needed. This principle is crucial for maintaining the operational continuity of organizations. Cyberattacks like Distributed Denial-of-Service (DDoS) attacks directly threaten availability by overwhelming systems with traffic, rendering services unavailable to legitimate users. One notable example is the 2016 Dyn DNS attack, which caused major disruptions to websites like Twitter and Netflix, highlighting how availability can be compromised on a large scale.[4]

These three principles should be balanced to ensure a secure and functional information system. As cybersecurity threats evolve, organizations should prioritize all three aspects of the CIA triad in their security strategies.

Authentication and Authorization

Authentication and authorization are two interrelated concepts that play a critical role in controlling access to information systems. Authentication is the process of verifying the identity of a user or system. In contrast, authorization determines what an authenticated user is allowed to do within a system.

1. **Authentication** verifies that a user is who they claim to be. Common methods include passwords, biometrics (such as fingerprints or facial recognition), and multi-factor authentication (MFA). MFA, in particular, has become a cornerstone of modern cybersecurity practices because it combines two or more authentication methods to enhance security. For example, many organizations now require both a password and a one-time code sent to a user's phone. This layered approach significantly reduces the likelihood of unauthorized access, as demonstrated by the effectiveness of MFA in preventing phishing attacks.[5]

2. **Authorization**, on the other hand, controls the actions a user can perform once authenticated. This process often involves assigning permissions based on a user's role within an organization. For instance, a system administrator might have more extensive privileges than a regular user, enabling them to perform tasks such as installing software or configuring system settings. Ensuring proper authorization is essential for preventing insider threats, where users exploit their access rights to compromise security.

A well-known case involving authorization failure is the Edward Snowden incident, where Snowden, a contractor for the NSA, misused his authorized access to leak sensitive government information.[6]

Encryption

Encryption is the process of converting plaintext into ciphertext. This makes data unreadable to unauthorized individuals. This method is widely used to protect the confidentiality of sensitive information during transmission and storage. Symmetric encryption and asymmetric encryption are two common types of encryption.

1. **Symmetric encryption** uses the same key to both encrypt and decrypt data. While it is fast and efficient, its major drawback is the need for a secure method to share the key between the sender and the recipient.

2. **Asymmetric encryption** uses a pair of keys; one public and one private. The public key is used to encrypt data, while the private key is used to decrypt it. This method is more secure, as the private key never has to be shared. One prominent use of asymmetric encryption is in SSL/TLS protocols, which secure web communications between users and websites[7].

Encryption is essential for maintaining confidentiality and protecting data from unauthorized access. In a recent survey, 76% of organizations reported using encryption to protect sensitive customer information, underscoring its widespread adoption.[8]

Firewalls and Intrusion Detection Systems (IDS)

Firewalls and Intrusion Detection Systems (IDS) are two key technologies used to protect networks from unauthorized access and cyber threats.

1. **Firewalls** act as a barrier between a trusted internal network and untrusted external networks, such as the internet. They monitor and control incoming and outgoing traffic based on predetermined security rules. Firewalls can be hardware-based, software-based, or cloud-based, depending on the organization's needs. For example, many enterprises deploy next-generation firewalls (NGFW), which integrate traditional firewall capabilities with advanced features such as deep packet inspection and threat intelligence.[9]

2. **Intrusion Detection Systems (IDS)**, on the other hand, monitor network traffic for suspicious activity and alert administrators if a potential threat is detected. IDS can be configured to monitor specific network segments or entire infrastructures. One notable example is the Snort IDS, an open-source tool widely used for real-time traffic analysis and packet logging. By deploying IDS alongside firewalls, organizations can enhance their ability to detect and respond to security incidents proactively.[10]

Zero Trust Model

The Zero Trust security model is a paradigm shift from traditional network security approaches. Instead of assuming that internal users and systems are trustworthy, Zero Trust operates on the principle of "never trust, always verify." This model requires continuous validation of every user and device attempting to access network resources, regardless of their location within the network.

The Zero Trust model gained widespread attention in response to the increasing sophistication of cyberattacks and insider threats. According to Forrester, 83% of organizations plan to adopt a Zero Trust strategy by 2025.[11] An example of a Zero Trust implementation is Google's BeyondCorp framework, which enables employees to work securely from any location without relying on a traditional virtual private network (VPN).[12]

Core Concepts of Cybersecurity

Understanding the key concepts and terminology of cybersecurity is crucial for defending against the wide array of threats that individuals, organizations, and governments face today. From the CIA triad to the Zero Trust model, these principles help shape the security strategies that safeguard sensitive information and ensure the availability of critical systems. As cybersecurity threats evolve, so too should our understanding of the underlying concepts that drive the field, ensuring that we remain one step ahead of those who seek to exploit vulnerabilities.

1.3 The CIA Triad: Confidentiality, Integrity, and Availability

The CIA triad: Confidentiality, Integrity, and Availability and explained earlier, is a fundamental model in cybersecurity. This framework is essential for understanding how to protect sensitive information and maintain the operational integrity of digital systems. Each element of the CIA triad represents a critical aspect of information security, and together, they form the cornerstone of modern cybersecurity practices. According to Eric Cole, a cybersecurity expert, "The CIA triad is a simple but comprehensive model that helps guide security professionals in balancing these three critical components."[1]

Confidentiality: Protecting Sensitive Information

Confidentiality involves ensuring that information is only accessible to authorized individuals. The goal of confidentiality is to prevent unauthorized users from accessing sensitive data, which could result in privacy violations, financial losses, or reputational damage. A common method for maintaining confidentiality is through encryption, which transforms plaintext into ciphertext, making data unreadable to anyone without the proper decryption key. For example, in the 2013 Yahoo data breach, the attackers gained access to encrypted passwords, but the company failed to maintain confidentiality due to weak encryption practices. This breach affected over three billion accounts, underscoring the importance of robust encryption technologies.[2]

The principle of confidentiality also applies to physical and digital access controls. Organizations often use multi-factor authentication (MFA), biometric systems, and role-based access control (RBAC) to ensure that only authorized personnel can access specific data. As cybersecurity expert Bruce Schneier puts it, "Confidentiality is not just about encryption; it's about ensuring that the right people have access to the right information."[3]

Furthermore, confidentiality extends to data classification policies that categorize information based on its sensitivity. For example, governments classify information as public, confidential, or top secret, depending on its level of sensitivity. By doing so, they ensure that the most critical information is afforded the highest level of protection.

According to the 2021 Global Information Security Workforce Study, 76% of organizations reported confidentiality breaches as their primary security concern, highlighting its critical importance in today's digital landscape.[4]

Integrity: Ensuring Data Accuracy and Trustworthiness

Integrity refers to the accuracy, consistency, and trustworthiness of data over its lifecycle. Maintaining data integrity means ensuring that information is not altered or tampered with by unauthorized parties. This is particularly important in industries like finance, healthcare, and government, where even minor alterations to data can lead to catastrophic outcomes. As William Stallings explains in his book *Cryptography and Network Security*, "Data integrity ensures that information remains in its original state and has not been altered by malicious actors or system errors."[5]

One prominent instance of an integrity breach was during the 2015 Bank of Bangladesh cyber robbery, in which attackers changed transaction data of the bank, therefore causing approximately $81 million stolen.[6] The SWIFT financial system's integrity was compromised by the hackers; hence, bogus transactions seemed to be valid. Such events highlight the vital need for integrity systems in financial systems. Organizations use hash functions that provide unique fixed-length outputs (hashes) for input data to guarantee data integrity. In has functions, if a single bit of the original data be altered, the hash value will vary and provide a simple means of detection of manipulation.

To ensure data integrity and authenticity during transmission, digital signatures and message authentication codes (MACs) also are applied.

Even when data flows between systems, these techniques help to preserve confidence in its accuracy.

Maintaining patient record integrity is absolutely vital in healthcare if safe and efficient treatment is to be given. An altered medical history of a patient could result in harmful therapeutic mistakes.

Health organizations thus make use of safe electronic health records (EHRs) systems with robust integrity safeguards to stop illegal modification of patient data.[7]

Availability: Ensuring Access to Information

Availability refers to ensuring that information and systems are accessible to authorized users whenever they are needed. It is not enough to simply protect the confidentiality and integrity of data; systems should also remain available to authorized users, especially during critical operations. A disruption in availability can lead to lost revenue, service interruptions, or even life-threatening situations, depending on the industry.

Availability can be threatened by several factors, including distributed denial-of-service (DDoS) attacks, hardware failures, and natural disasters. One of the most notable examples of a DDoS attack was the 2016 Dyn DNS attack, which affected major websites like Amazon, Twitter, and Netflix by overwhelming the DNS provider's servers with traffic. This attack highlighted the fragility of critical infrastructure to availability attacks by interrupting access to many websites for hours [8].

Organizations use geographically separated data centers, data replication, and backup power sources among other redundancy and fault-tolerance strategies to guarantee availability. High availability systems are used in important sectors such banking and healthcare to guarantee that services continue to run even during unplanned failures. Cybersecurity consultant Marcus Ranum points out, "Availability is often the most overlooked aspect of the CIA triad, but it is just as crucial as confidentiality and integrity."[9]

In the cloud computing era, ensuring availability also involves managing the risks associated with service-level agreements (SLAs) and provider outages. Many businesses rely on cloud services for their day-to-day operations, so any disruption in cloud availability can have widespread consequences. The Amazon Web Services (AWS) outage in 2020, which affected thousands of websites and online services, is an example of how cloud availability issues can impact businesses globally.[10]

Balancing the CIA Triad in Practice

In practice, cybersecurity professionals should balance confidentiality, integrity, and availability to create secure and functional systems. Over emphasizing one aspect of the CIA triad can lead to weaknesses in others. For example, if a company only cares about confidentiality and integrity, they might forget about availability, which could make systems hard to get to in an emergency. On the other hand, putting availability ahead of privacy can cause holes that let private data get out.

To make a good defense plan, you need to look at the whole picture and treat all three parts of the CIA triad equally. Keeping this balance will become more important as businesses use cloud computing, AI-driven security, and quantum encryption. As the 2021 Verizon Data Breach Investigations Report indicates, 95% of breaches can be attributed to failures in one or more aspects of the CIA triad, underscoring its critical role in preventing cyberattacks [11].

The Importance of the CIA Triad

The CIA triad, Confidentiality, Integrity, and Availability, remains the foundation of cybersecurity, guiding the development of security policies, technologies, and strategies. By focusing on these three critical elements, organizations can protect sensitive data, ensure the accuracy of information, and maintain the operational availability of their systems. As cyber threats continue to evolve, the CIA triad will remain a crucial framework for defending against both current and future attacks.

1.4 Types of Cyber Threats and Attacks

Cybersecurity threats and attacks have become increasingly sophisticated over time. Understanding the types of threats organizations and individuals face is vital for developing effective defensive measures. This section outlines the most common types of cyber threats and attacks, including examples and their potential impact.

Malware: A Multi-Faceted Cyber Threat

Malware, or malicious software, encompasses a broad range of harmful programs such as viruses, worms, Trojans, ransomware, and spyware. Malware is often designed to infiltrate or damage systems without the user's knowledge or consent. One of the most infamous malware attacks occurred in 2017 with the WannaCry ransomware outbreak, which affected over 200,000 computers across 150 countries. The ransomware encrypted files and demanded payment in Bitcoin for decryption keys, highlighting the economic and operational damage malware can caus[1]. According to Saini et al., "malware remains one of the most prevalent and dangerous forms of cyber threats due to its evolving nature."[2]

Viruses and worms are among the oldest forms of malware. A virus requires a host file or application to spread, while a worm can replicate and spread on its own. Worms like the Morris Worm of 1988 demonstrated the potential for widespread damage, even in the early days of the internet. Worms have since evolved, and modern versions, like Conficker, have targeted millions of machines globally.

Trojan horses masquerade as legitimate software, tricking users into installing them. Once installed, they can create backdoors for unauthorized access or steal sensitive data. In 2013, the Zeus Trojan was responsible for stealing banking credentials and siphoning millions of dollars from financial institutions worldwide.[3]

Ransomware has become a significant threat to both individuals and organizations. This type of malware encrypts the victim's data and demands payment for its release. For example, the Colonial Pipeline ransomware attack in 2021 disrupted fuel supply across the eastern United States, resulting in a ransom payment of $4.4 million.[4]

Phishing and Social Engineering Attacks

Phishing is a social engineering attack that involves tricking individuals into divulging personal or confidential information by posing as a trustworthy entity, often through email or websites. Spear-phishing, a more targeted version of phishing, focuses on specific individuals or organizations, often using detailed information to increase credibility. In 2016, the Democratic National Committee (DNC) was a victim of a spear-phishing attack that led to a significant data breach during the US presidential election campaign.[5]

According to Uma and Padmavathi, phishing attacks "rely on the manipulation of human behavior rather than technical vulnerabilities, making them difficult to prevent."[6] Attackers may pose as banks, government agencies, or trusted service providers to trick users into revealing sensitive information such as passwords, credit card numbers, or Social Security numbers.

Business Email Compromise (BEC), a type of phishing attack that targets corporate executives and finance departments. Attackers impersonate a high-ranking employee and instruct the recipient to transfer funds or provide sensitive financial information. The FBI reported that BEC scams led to global losses of over $1.8 billion in 2020 alone.[7]

Denial-of-Service (DoS) and Distributed Denial-of-Service (DDoS) Attacks

A Denial-of-Service (DoS) attack aims to make a website or network unavailable by overwhelming it with traffic or exploiting vulnerabilities. A Distributed Denial-of-Service (DDoS) attack is a more advanced form of this attack, where traffic is generated from multiple sources, often through a botnet (a network of compromised computers). One of the largest DDoS attacks occurred in 2016, targeting Dyn, a major Domain Name System (DNS) provider, and disrupting services for major websites such as Twitter, Netflix, and Amazon.[8]

Aslan et al. explain that "DDoS attacks have evolved in complexity, leveraging IoT devices and cloud services to amplify their impact."[9]

Attackers often target critical infrastructure, as was the case in the 2020 Akamai attack, which reached a peak of over 1.44 terabits per second. This was one of the largest recorded at the time.

Advanced Persistent Threats (APTs)

Advanced Persistent Threats (APTs) are long-term targeted attacks carried out by well-funded and highly skilled adversaries, often state-sponsored or associated with organized cybercrime groups. APTs aim to infiltrate networks and remain undetected for extended periods to exfiltrate sensitive data, gather intelligence, or sabotage systems.

One of the most notable APTs is the Stuxnet worm, discovered in 2010, which targeted Iran's nuclear facilities. Stuxnet was a sophisticated cyber-weapon designed to disrupt centrifuges used for uranium enrichment. This highlights the potential for APTs to cause real-world physical damage.[10]

Agrafiotis et al. explain, "APTs are characterized by their stealth, persistence, and ability to evade traditional cybersecurity defenses."[11] These attacks are often attributed to nation-state actors engaging in cyber espionage or cyber warfare.

Insider Threats

Insider threats arise when an employee, contractor, or business partner with authorized access to systems misuses their privileges to steal or sabotage data. Insider threats can be intentional, such as data theft, or unintentional, where an employee unknowingly introduces vulnerabilities by violating security protocols. The Edward Snowden case is one of the most well-known examples of an insider threat. Snowden, a former NSA contractor, leaked classified government data in 2013. This exposed massive government surveillance operations.[12]

Liu et al. argue, "Insider threats pose unique challenges because they originate from trusted individuals with legitimate access, making detection difficult."[13] Organizations often implement behavioral monitoring and privileged access management (PAM) systems to mitigate insider risks.

Man-in-the-Middle (MitM) Attacks

A Man-in-the-Middle (MitM) attack occurs when an attacker intercepts and potentially alters communication between two parties without their knowledge. MitM attacks often target users on unsecured public Wi-Fi networks or exploit vulnerabilities in websites that do not use encryption. One infamous MitM attack was the FREAK vulnerability in 2015, which affected HTTPS connections, exposing millions of users to potential eavesdropping.[14]

Humayun et al. emphasize that "MitM attacks can compromise confidentiality and integrity by allowing attackers to alter or steal sensitive data during transmission."[15] Using end-to-end encryption and Virtual Private Networks (VPNs) are common methods for mitigating MitM risks.

Zero-Day Exploits

Zero-day exploits target software vulnerabilities that are unknown to the vendor or have not yet been patched. These attacks are particularly dangerous because they occur before developers have a chance to fix the vulnerability. Zero-day vulnerabilities can remain undiscovered for years, allowing attackers to exploit them without detection. For instance, the Microsoft Exchange Server zero-day vulnerabilities discovered in 2021 were exploited by attackers to gain unauthorized access to email accounts and sensitive data.[16]

According to a study by Gunduz and Das, "zero-day vulnerabilities represent one of the greatest challenges in cybersecurity due to their unpredictability and potential to cause significant harm before detection"[17]. Cybersecurity professionals often use intrusion detection systems (IDS) and machine learning models to detect anomalies that may indicate zero-day exploits.

The Evolving Cyber Threat Landscape

Cyber threats and attacks continue to evolve, driven by advancements in technology and the increasing sophistication of attackers. From traditional malware and phishing to complex APTs and zero-day exploits, organizations should remain vigilant and adopt a comprehensive, multi-layered approach to cybersecurity. As cybercriminals develop new attack methods, it is crucial for organizations to stay informed about emerging threats and implement proactive defenses to protect their systems and data.

1.5 Basic Principles of Digital Protection

In today's hyperconnected world, digital protection is essential for safeguarding data, systems, and networks from cyber threats. The basic principles of digital protection are designed to prevent unauthorized access, ensure data integrity, and maintain the availability of information. These principles are vital for individuals, organizations, and governments to mitigate risks posed by cyberattacks. Understanding the core tenets of digital security helps in formulating effective strategies to protect digital assets. We have previously dealt extensively with Confidentiality, Integrity, Availability, Authentication, and Authorization, which are the tenets of guarding our digital assets. Now, let's look at non-repudiation, defense in depth, and security awareness training in details.

Non-repudiation: Accountability in Digital Transactions

Non-repudiation is the assurance that someone cannot deny the authenticity of their signature on a document or a message sent. This principle ensures that actions taken in digital systems are traceable and verifiable. Digital signatures and encryption algorithms are the tools typically employed to establish non-repudiation.

For example, in e-commerce, non-repudiation ensures that once a buyer agrees to purchase a product and the transaction is confirmed, neither the buyer nor the seller can deny the exchange. According to Jouini et al., "Non-repudiation is critical for maintaining trust in digital transactions, especially in industries such as banking and finance, where accountability is paramount."[6]

Defense in Depth: Layered Security Strategy

Defense in depth is a layered security strategy that employs multiple protective measures to safeguard information. The idea is that if one layer of defense fails, others will still provide protection. This principle is vital in modern cybersecurity because it addresses the complexity and sophistication of today's threats. Defense in depth includes a combination of physical security, network security, endpoint protection, and data encryption.

An example of defense in depth is using firewalls, intrusion detection systems (IDS), antivirus software, and data encryption concurrently. Each layer serves a distinct purpose, enhancing the overall security posture of the organization. According to Gunduz and Das, "Layered security approaches are crucial in mitigating risks, particularly in sectors that face a high level of targeted cyberattacks." [7]

Security Awareness and Training

Another essential principle of digital protection is security awareness and training. Cybersecurity is not just a technical issue but a human one as well. Organizations should educate employees on best practices, such as identifying phishing emails, using strong passwords, and following security protocols. Conduct regular training sessions ensure that employees are aware of the latest threats and how to respond to them.

Uma and Padmavathi highlight that "The human element is often the weakest link in cybersecurity, and robust security awareness programs are essential to prevent inadvertent security breaches." [8]

The Importance of Basic Principles

The basic principles of digital protection, confidentiality, integrity, availability, authentication, authorization, non-repudiation, defense in depth, and security awareness, form the foundation of a strong cybersecurity strategy. As cyber threats evolve, organizations should continuously adapt their security measures to protect their digital assets effectively. Understanding and implementing these principles is critical in maintaining a secure and resilient digital environment.

1.6 The Cybersecurity Ecosystem: Stakeholders and Their Roles

The cybersecurity ecosystem is an intricate web of stakeholders who work together to protect the integrity, confidentiality, and availability of information and systems. Each stakeholder plays a unique role in ensuring that digital assets are safeguarded from the ever-evolving threat landscape. From individuals and private corporations to government entities and international organizations, the collaborative efforts of these actors are essential to maintaining a resilient cybersecurity infrastructure.

Individuals: The First Line of Defense

Individuals are often considered the first line of defense in the cybersecurity ecosystem. Personal responsibility in managing digital security has become increasingly important as individuals now handle vast amounts of sensitive data, both personally and professionally. The rise of phishing attacks, where malicious actors trick users into providing sensitive information, illustrates how individuals are frequently targeted by cybercriminals. A study by IBM found that "human error is the primary cause of 95% of cybersecurity breaches." [1]

To mitigate these risks, individuals should adopt cyber hygiene practices, such as using strong passwords, enabling multi-factor authentication, and avoiding suspicious emails or links. Training and awareness programs provided by employers and educational institutions are critical in equipping individuals with the knowledge needed to identify potential threats.

Private Sector Organizations: Guardians of Data and Infrastructure

Private sector organizations are major stakeholders in the cybersecurity ecosystem, responsible for protecting their data and ensuring the integrity of their digital systems. With the digital transformation of businesses, organizations now face significant cybersecurity risks that can lead to financial losses, damaged reputation, and regulatory penalties. According to Verizon's 2021 Data Breach Investigations Report, 43% of cyberattacks target small businesses, underscoring the vulnerability of organizations of all sizes. [2]

Corporations are responsible for implementing robust cybersecurity frameworks, including firewalls, encryption, and endpoint protection solutions. In addition, many companies are adopting zero trust models, which require continuous verification of users and devices, regardless of whether they are inside or outside the organization's network. As Richard Clarke, a cybersecurity advisor, has emphasized, "Corporate responsibility in cybersecurity is not just about compliance; it's about survival in a digital-first economy."[3]

Furthermore, industries such as healthcare and finance, which handle sensitive personal and financial information, should comply with strict cybersecurity regulations, including the Health Insurance Portability and Accountability Act (HIPAA) and the General Data Protection Regulation (GDPR). Failure to comply can result in significant penalties, as seen in the Equifax breach of 2017, where the company was fined $700 million for failing to protect the personal data of 147 million customers. [4]

Government Agencies: Regulatory and Defense Roles

Government agencies play a dual role in the cybersecurity ecosystem: they are both defenders of national infrastructure and regulators of cybersecurity policies. On one hand, governments are responsible for protecting critical infrastructure such as power grids, financial systems, and healthcare services from cyberattacks. For example, the Cybersecurity and Infrastructure Security Agency (CISA) in the United States coordinates national efforts to defend against cyber threats, working closely with both the public and private sectors.[5]

On the regulatory side, governments set cybersecurity standards and guidelines that organizations should follow. For instance, the National Institute of Standards and Technology (NIST) provides a cybersecurity framework that many organizations adopt as a baseline for their security practices. Governments also legislate privacy and data protection laws, such as GDPR in the European Union, which sets strict standards for how organizations collect, store, and use personal data.

In recent years, nation-state attacks have become a growing concern for governments as foreign actors attempt to infiltrate critical infrastructure and steal sensitive information. The 2020 SolarWinds attack, attributed to Russian hackers, compromised numerous US government agencies. This highlights the vulnerability of even the most secure systems [6]. As cybersecurity threats from foreign governments increase, nations should invest heavily in both offensive and defensive cybersecurity capabilities.

Cybersecurity Firms: The Frontline Protectors

Cybersecurity firms are the frontline protectors in the battle against cybercrime. These organizations develop and provide tools, services, and technologies designed to detect, prevent, and mitigate cyber threats. The global cybersecurity industry is growing rapidly, with projections indicating it will reach $400 billion by 2026.[7]

Leading firms such as Palo Alto Networks, Symantec, and McAfee offer a wide range of products, including firewalls, antivirus software, and advanced threat detection systems. In addition, these firms provide incident response services, helping organizations recover from cyberattacks by identifying the source of breaches and implementing measures to prevent future attacks. Cybersecurity companies also offer managed security services to organizations that lack the resources to maintain their own in-house security teams.

Beyond technology, cybersecurity firms contribute to the ecosystem by conducting research on emerging threats and vulnerabilities. The Unit 42 research team at Palo Alto Networks, for example, provides critical insights into new malware strains and attack tactics. This kind of research helps organizations stay ahead of cybercriminals.[8]

Law Enforcement: Investigating and Prosecuting Cybercriminals

Law enforcement agencies, both at the national and international levels, are tasked with investigating cybercrimes and prosecuting cybercriminals. The complexity of cybercrimes often transcends national borders, requiring cooperation between multiple jurisdictions. Organizations such as INTERPOL and Europol play vital roles in facilitating international cooperation to combat cybercrimes, including ransomware attacks, identity theft, and fraud.

In 2020, Europol and the FBI worked together in Operation DisrupTor, which dismantled a large-scale dark web marketplace that facilitated the sale of illegal drugs, weapons, and stolen data.[9] This collaboration highlights the importance of cross-border cooperation in addressing the global nature of cybercrime. Additionally, law enforcement agencies increasingly rely on digital forensics to trace the origins of cyberattacks and gather evidence for prosecution.

International Organizations: Facilitating Global Cooperation

In an interconnected world, international organizations are essential in facilitating global cooperation on cybersecurity issues. The United Nations (UN), through its International Telecommunication Union (ITU), plays a key role in establishing global cybersecurity frameworks and promoting the sharing of best practices among member states.[10] Similarly, the North Atlantic Treaty Organization (NATO) focuses on defending member countries against cyber threats, emphasizing collective defense in the face of cyberattacks.

International organizations also provide platforms for dialogue on cybersecurity policy, standards, and governance. For example, the Global Forum on Cyber Expertise (GFCE) brings together governments, the private sector, and civil society to build cybersecurity capacity worldwide, particularly in developing countries.[11]

As cyber threats continue to evolve, international collaboration will be crucial in addressing challenges such as cyber espionage, intellectual property theft, and cyber warfare.

The Importance of Collaboration in Cybersecurity

Cybersecurity is no one person's, company's, or government's only responsibility in today's interconnected digital world. All of the participants in the cybersecurity ecosystem should actively participate and coordinate their efforts. Each of the people, companies, governments, academics, international organizations, and complex network members,who together help to protect digital infrastructure and preserve information system security, plays a vital part.

Cooperation among these several organizations is not only valuable but also quite necessary. Practicing basic cyber hygiene, that is, using strong passwords and enabling multi-factor authentication, at the individual level comes first in defense. To reduce vulnerabilities, companies should commit to strong cybersecurity systems, staff training, and quick incident response procedures. While governments should build alliances with the private sector to improve defense capability, they also have to create and enforce rules to guarantee compliance with cybersecurity criteria. Globally, international cooperation is absolutely critical. Nations should share intelligence, align legal systems, and cooperate on threat reduction strategies since cyberattacks are borderless, usually starting in one country but aiming at another. Initiatives, including cross-border information-sharing platforms, NATO's cyber defense policies, and the GDPR of the European Union, show group efforts in confronting shared concerns. A coordinated multi-stakeholder approach becomes increasingly important as cyberattacks become more sophisticated and widespread. Companies, governments, and people have to break out from their silos, share knowledge, and cooperate to create strong defenses. Working together guarantees the pooling of resources, knowledge, and technologies, guaranteeing a proactive reaction to the always-changing terrain of cyber threats. The strength of our security in this shared digital environment resides in the strength of our relationships.

Test Your Knowledge on Foundations of Cybersecurity

Q1: What is the main purpose of cybersecurity in the digital world?

Q2: How has cybersecurity evolved over time?

Q3: What are the key components of the CIA triad in cybersecurity?

Q4: What is the significance of the CIA triad?

Q5: How do cyber threats differ from traditional security threats?

Q6: What are some of the common types of cyberattacks?

Q7: How do social engineering attacks manipulate human behavior?

Q8: What role does encryption play in digital protection?

Q9: How does multi-factor authentication enhance security?

Q10: What is the role of firewalls in cybersecurity?

Q11: Why is privileged access management important in cybersecurity?

Q12: How do businesses implement incident response plans?

Q13: What is the role of cybersecurity frameworks like NIST?

Q14: What are zero-day exploits, and why are they dangerous?

Q15: How does risk management help in mitigating cybersecurity threats?

ANSWERS

Q1: A: The primary purpose of cybersecurity is to protect information systems, networks, and data from unauthorized access, theft, damage, and other cyber threats.

Q2: A: Cybersecurity has evolved from simple password protection and antivirus software to more advanced defense mechanisms, including encryption, multi-factor authentication, AI-driven threat detection, and proactive threat hunting.

Q3: A: The CIA triad includes Confidentiality, Integrity, and Availability, which are foundational to protecting data and ensuring its proper use.

Q4: A: The CIA triad ensures that data remains secure by preventing unauthorized access (Confidentiality), protecting its accuracy (Integrity), and making it accessible to authorized users when needed (Availability).

Q5: A: Cyber threats are constantly evolving and more sophisticated and can target multiple systems simultaneously, unlike traditional threats, which may be more physical or location-based.

Q6: A: Common types include phishing, malware, ransomware, social engineering, and denial-of-service attacks.

Q7: A: Social engineering attacks deceive individuals into divulging confidential information by exploiting psychological tactics, such as impersonation or urgency.

Q8: A: Encryption transforms data into unreadable formats, preventing unauthorized access and maintaining the integrity and confidentiality of sensitive information.

Q9: A: Multi-factor authentication requires multiple forms of verification (e.g., passwords and biometrics), adding an additional layer of security.

Q10: A: Firewalls act as barriers between trusted internal and untrusted external networks, filtering out malicious traffic.

Q11: A: It helps ensure that only authorized users can access sensitive systems, minimizing the risk of insider threats and data breaches.

Q12: A: Incident response plans outline specific steps for detecting, responding to, and recovering from cyber incidents, ensuring minimal disruption to operations.

Q13: A: Frameworks provide organizations with structured guidelines for assessing and improving their security posture, ensuring compliance with industry standards.

Q14: A: Zero-day exploits target vulnerabilities that have not yet been patched, making them highly dangerous as there are no immediate defenses.

Q15: A: Risk management involves identifying potential threats, assessing their impact, and implementing strategies to minimize or eliminate vulnerabilities.

CHAPTER 2

TECHNOLOGICAL INNOVATIONS IN CYBERSECURITY

Note: All references in this chapter are on page 395 - 396

This chapter talks about the cutting-edge technologies that are changing the way cybersecurity is done and will change the way digital defense grows in the future. As technology improves, so do the ways that attackers and guards try to trick each other. We will talk about how AI and machine learning are changing how risks are found and making it easier for systems to find problems and attacks more quickly than ever. The fact that blockchain technology was first created for cryptocurrencies shows how strong it is as a way to keep deals and data safe. Although it causes problems for current cryptographic standards, quantum cryptography, a very advanced form of encryption based on quantum physics, could provide security that can't be broken.

As more businesses use cloud-based tools, this part also talks about how cloud safety has upgraded. We'll also talk about end-to-end encryption and how biometrics are becoming more important for identification. By the end of this chapter, you'll know how these new technologies are both making defenses stronger and adding new threats. This chapter also talks about how hard it is to stay ahead of cybercriminals, who are always finding new ways to use technology to start more complex attacks.

2.1 Artificial Intelligence and Machine Learning in Threat Detection

The rise of Artificial Intelligence (AI) and Machine Learning (ML) has fundamentally transformed the field of cybersecurity. These technologies allow systems to detect, analyze, and respond to cyber threats with greater speed and accuracy than traditional methods. AI and ML are now indispensable tools in the fight against evolving cyber threats, as they provide the ability to process vast amounts of data, identify anomalies, and predict future attacks based on patterns.

The Role of AI in Threat Detection

AI in cybersecurity is primarily used to automate threat detection and response processes. Unlike traditional security systems, which rely on predefined rules and signature-based detection, AI systems can adapt and learn from the data they process. AI models analyze traffic patterns, detect irregular behaviors, and flag potential threats in real time. This capability is particularly valuable in identifying zero-day vulnerabilities, where malicious actors exploit unknown weaknesses in software before they can be patched.

According to a study by Mohammadi et al., "AI-powered threat detection systems offer faster and more accurate threat detection by analyzing large datasets and identifying patterns that traditional systems might miss." [1] AI's ability to sift through vast amounts of data and recognize patterns has become essential in defending against modern cyber threats.

One notable application of AI in threat detection is in intrusion detection systems (IDS), which use AI algorithms to monitor network traffic and identify abnormal behavior. For example, Darktrace, a cybersecurity company, leverages AI to provide real-time threat detection using machine learning algorithms that can detect even subtle deviations from normal behavior across networks.[2] Darktrace's AI-driven platform continuously learns from an organization's network behavior, making it highly effective at detecting sophisticated and previously unseen attacks.

Machine Learning: Enhancing Predictive Capabilities

Machine Learning (ML) plays a critical role in improving cybersecurity by learning from historical data and predicting future cyberattacks. ML models are trained on large datasets of both legitimate and malicious activities, enabling them to distinguish between normal and abnormal behaviors. Unlike static security systems, ML algorithms evolve as they process more data, becoming more accurate over time in detecting threats.

For example, supervised learning algorithms can be trained to classify different types of network traffic as either safe or malicious. These models use labeled data to learn the characteristics of various threats, such as malware or phishing attempts. Once trained, the algorithm can detect similar threats in the future with high accuracy. On the other hand, unsupervised learning algorithms do not rely on labeled data but instead identify anomalies in network behavior. This makes them particularly useful for identifying unknown or emerging threats.

As Chio and Freeman explain, "ML systems excel at detecting unknown threats because they can identify patterns that humans might overlook, especially in large and complex datasets."[3] This predictive capability is crucial in a world where new types of malware and cyberattacks emerge daily. ML algorithms can adapt to these changes and offer a dynamic defense against evolving threats.

One example of ML in action is the use of neural networks to detect malware. Neural networks, which mimic the human brain's structure and learning process, can analyze data from multiple sources and detect subtle changes in behavior that may indicate a malware infection. These systems are particularly effective at detecting fileless malware, which is difficult for traditional antivirus software to catch because it operates in a system's memory rather than on disk.[4]

AI and ML in Combating Phishing Attacks

Phishing attacks remain one of the most common methods cybercriminals use to steal sensitive information. AI and ML technologies are increasingly being used to combat phishing by detecting malicious emails and websites before they reach the end user. These technologies analyze email content, URLs, and sender metadata to identify phishing attempts based on known characteristics of such attacks.

AI systems can detect phishing attacks by analyzing language patterns and comparing them to a database of known phishing attempts. For instance, Google's Gmail uses ML algorithms to block more than 100 million phishing emails per day by recognizing suspicious email patterns and automatically filtering them out[5]. Additionally, these systems continuously learn from new phishing emails, improving their ability to detect future attacks.

ML algorithms also play a critical role in detecting spear-phishing attacks, which target specific individuals or organizations. Unlike traditional phishing attacks, spear-phishing involves highly personalized emails, making them harder to detect. By analyzing past email behavior and identifying discrepancies, ML algorithms can flag suspicious emails and prevent them from reaching the intended recipient.

Real-Time Threat Detection and Response

One of the most significant advantages of AI and ML in cybersecurity is the ability to detect and respond to threats in real time. Traditional security systems often react to attacks after they have occurred, whereas AI systems can proactively detect and mitigate threats before they cause significant damage. This capability is essential in modern cybersecurity, where attackers can compromise systems in a matter of seconds.

AI-powered Security Information and Event Management (SIEM) systems are an example of real-time threat detection and response. These systems collect and analyze security data from across an organization's network, using AI algorithms to detect anomalies and respond to potential threats in real time. When a threat is detected, the SIEM system can automatically initiate a response, such as isolating affected devices or blocking malicious traffic.

For example, IBM's QRadar SIEM platform uses AI to analyze network traffic and log data, detect suspicious activity, and trigger automated responses to mitigate potential threats.[6] This reduces the time it takes for security teams to respond to incidents and minimizes the impact of cyberattacks.

As Alazab et al. note, "Real-time threat detection enabled by AI significantly reduces the response time to cyber incidents, thereby limiting potential damage."[7] This proactive approach to cybersecurity is becoming increasingly necessary as cyberattacks become more frequent and sophisticated.

Challenges and Limitations of AI and ML in Cybersecurity

Despite the many advantages of AI and ML in threat detection, these technologies are not without limitations. One significant challenge is the potential for false positives, where legitimate activity is mistakenly flagged as malicious. This can overwhelm security teams with unnecessary alerts and hinder their ability to respond to genuine threats.

Additionally, cybercriminals are now using AI and ML to develop more advanced attacks. For instance, adversarial machine learning involves manipulating ML models by feeding them incorrect or deceptive data to trick them into making incorrect decisions. Attackers can exploit these vulnerabilities to bypass security systems that rely on AI and ML.

As West and Bhattacharya explain, "While AI and ML are powerful tools for threat detection, they are not immune to manipulation by cyber adversaries."[8] To address these challenges, cybersecurity professionals should continuously refine AI and ML models and incorporate human oversight into the decision-making process.

Conclusion

AI and ML have revolutionized the field of cybersecurity by enhancing the speed, accuracy, and predictive capabilities of threat detection systems. These technologies enable organizations to detect and respond to cyber threats in real-time, providing a critical defense against an increasingly sophisticated threat landscape. As cyberattacks continue to evolve, AI and ML will play an even more significant role in maintaining the security and integrity of digital systems.

2.2 Blockchain Technology and Its Security Applications

Blockchain technology has emerged as one of the most transformative innovations in the realm of digital security. Originally developed to support cryptocurrencies such as Bitcoin, Blockchain technology is now being applied to various sectors for its ability to provide transparency, immutability, and decentralized security. Blockchain is considered particularly valuable in cybersecurity because it offers a robust method for securing data and transactions in distributed networks, where centralized control is neither feasible nor desirable.

The Fundamentals of Blockchain Technology

At its core, Blockchain is a decentralized digital ledger that records transactions across a network of computers. This ledger is divided into blocks, each containing a list of transactions. Once a block is completed, it is added to the chain in a linear, chronological order. Each block is linked to the previous one using a cryptographic hash, ensuring the immutability of the recorded data.

The decentralized nature of Blockchain ensures that no single entity controls the entire network. Instead, every participant in the network has access to a copy of the ledger, and changes can only be made through consensus mechanisms such as proof of work (PoW) or proof of stake (PoS). According to Mougayar, "Blockchain's decentralized architecture eliminates the need for trust between parties, as the system itself guarantees data integrity."[1]

Immutability: Ensuring Data Integrity and Security

One of Blockchain's most significant contributions to security is its ability to guarantee the immutability of data. Once a transaction is recorded on the Blockchain, it is virtually impossible to alter without altering every subsequent block, which would require the consensus of the entire network. This feature makes Blockchain highly resistant to tampering and fraud.

For example, Blockchain's immutability is particularly valuable in preventing data breaches and ensuring the integrity of records. In traditional databases, a malicious actor with access to the system could modify or delete data without detection. However, on a Blockchain, such alterations would be immediately visible, making unauthorized modifications extremely difficult. As Crosby et al. note, "Blockchain's immutability creates a secure environment for storing sensitive data, as unauthorized changes are virtually impossible."[2]

Blockchain technology is especially useful in sectors such as finance, healthcare, and supply chain management, where data integrity is critical. For instance, Blockchain is being explored as a solution for maintaining the security and accuracy of electronic health records (EHRs). By storing patient records on a Blockchain, healthcare providers can ensure that records remain unaltered while still allowing authorized access to medical professionals.[3]

Decentralization: Reducing Single Points of Failure

The decentralized nature of Blockchain makes it inherently secure against certain types of cyberattacks, particularly those that exploit single points of failure. In traditional centralized systems, if a hacker successfully breaches the central server, they gain access to all the system's data. However, in a Blockchain network, the data is distributed across multiple nodes, making it far more difficult for attackers to compromise the entire system.

For example, decentralized applications (dApps) built on Blockchain are less vulnerable to distributed denial-of-service (DDoS) attacks because there is no central server to overwhelm. In traditional systems, DDoS attacks can incapacitate entire networks by flooding them with traffic, but Blockchain-based networks are more resilient due to their distributed architecture [4]. In this way, Blockchain contributes to network resilience and reduce the risk of catastrophic failures.

According to De Filippi and Wright, "decentralized systems, such as Blockchain, inherently mitigate the risks associated with central points of failure, enhancing overall system security."[5]

Blockchain in Identity Management and Authentication

Blockchain technology has significant implications for identity management and authentication. Traditional identity verification systems often rely on centralized databases, which are vulnerable to breaches and misuse. Blockchain offers a decentralized solution where individuals can have more control over their personal information and share it only with trusted parties.

A key concept in this area is Self-Sovereign Identity (SSI), where individuals own and control their digital identities rather than relying on a third-party intermediary. Blockchain enables the creation of a distributed identity system in which users can store their identity credentials on the Blockchain and share them with service providers without exposing the entirety of their personal data. This process not only enhances privacy but also reduces the risk of identity theft.

For instance, Civic, a Blockchain-based identity verification platform, allows users to create and manage their own identities using Blockchain. The platform provides secure, decentralized verification of identity credentials, reducing the reliance on centralized identity management systems and preventing unauthorized access.[6]

Similarly, Microsoft's Azure Active Directory (AD) is experimenting with Blockchain for decentralized identity management, allowing individuals to control and share their data without intermediaries.[7] This shift towards decentralized identity management offers increased security and privacy for users while reducing the vulnerability of centralized identity repositories to breaches.

Blockchain in Supply Chain Security

Blockchain's ability to provide transparency and traceability is being applied in supply chain security, where the authenticity and integrity of products should be verified at each stage of the supply chain. Blockchain can create a transparent, tamper-proof record of each transaction, ensuring that every stakeholder in the supply chain can verify the provenance of goods and detect any attempts to tamper with or counterfeit products.

For example, IBM's Food Trust uses Blockchain to trace the journey of food products from farm to table. By recording every transaction on the Blockchain, stakeholders can verify the authenticity and safety of food products. This reduces the risks associated with food fraud, contamination, or tampering.[8] The application of Blockchain in the supply chain is especially important in industries such as pharmaceuticals, where counterfeit drugs can have life-threatening consequences.

The ability to trace goods through the entire supply chain also enhances accountability and compliance with industry regulations. As Tapscott and Tapscott explain, "Blockchain's transparency and immutability make it an ideal solution for ensuring product integrity in complex supply chains." [9]

Blockchain and Smart Contracts: Automating Security Processes

Smart contracts, which are self-executing contracts with the terms of the agreement written into code, are another significant security application of Blockchain. These contracts automatically enforce the terms of an agreement and reduce the need for intermediaries and increase security by eliminating human error or manipulation.

Smart contracts are particularly useful in financial transactions, where they can automatically execute payments once certain conditions are met. For example, Ethereum, a leading Blockchain platform, supports smart contracts that enable secure, automated transactions without the need for intermediaries such as banks. These contracts not only increase efficiency but also enhance security by ensuring that all parties adhere to the agreed-upon terms.[10]

Additionally, smart contracts can be used to automate compliance with regulatory requirements. In industries such as insurance, smart contracts can automatically enforce policies and claims. This reduces the risk of fraud and increase trust between parties. As Buterin, the creator of Ethereum, has stated, "Smart contracts on the Blockchain provide a reliable, tamper-proof way to enforce agreements and reduce the reliance on third-party intermediaries."[11]

Challenges and Limitations of Blockchain in Cybersecurity

Despite its many advantages, Blockchain is not without its challenges. One significant issue is scalability. As the size of the Blockchain grows, the amount of computing power required to process transactions increases, potentially leading to slower transaction times and higher costs. Additionally, while Blockchain is highly secure, it is not immune to attacks. For example, a 51% attack occurs when a group of miners gains control of more than half of the network's computing power, allowing them to manipulate the Blockchain.[12]

Moreover, the implementation of Blockchain in cybersecurity requires careful consideration of privacy. While Blockchain provides transparency, it can also expose sensitive information if not properly managed. Thus, the balance between transparency and privacy remains a key challenge in the deployment of Blockchain solutions.

Conclusion

Blockchain technology offers numerous security applications that enhance data integrity, transparency, and resilience. Its decentralized nature and immutability make it an ideal solution for industries where security and trust are paramount. From identity management to supply chain security, Blockchain's potential to transform cybersecurity is vast.

However, as with any technology, it should be implemented carefully, with consideration given to its limitations and potential vulnerabilities. As Blockchain continues to evolve, it will undoubtedly play a pivotal role in shaping the future of digital security.

2.3 Cloud Security Advancements

Cloud computing offers businesses flexibility, scalability, and cost savings for business and other organizations. However, this also presents unique security challenges. The security of data stored and processed in the cloud-based server has become a critical concern, as more organizations transition to cloud-based services. Over the past decade, significant advancements have been made in cloud security, addressing concerns related to data privacy, compliance, and protection from cyberattacks. These innovations are vital in maintaining the integrity and security of sensitive information in the cloud.

The Shift to Cloud Computing and Its Security Implications

Cloud computing enables organizations to store and process data over the Internet rather than relying on local servers or personal computers. This shift has accelerated in recent years due to the increasing demand for scalable infrastructure that supports remote work and data-intensive applications. However, the move to cloud services has introduced new security risks, such as data breaches, misconfigurations, and insider threats. According to Gartner, "By 2025, 99% of cloud security failures will be the customer's fault, primarily due to misconfigurations."[1]

These misconfigurations often occur when companies fail to properly secure cloud environments, leaving data exposed. For example, the Capital One data breach in 2019 resulted from a misconfigured firewall, allowing an attacker to access sensitive customer data stored in Amazon Web Services (AWS) cloud.[2] This breach underscores the importance of robust cloud security practices and the need for continuous monitoring.

Encryption: Protecting Data in Transit and at Rest

Encryption is one of the foundational technologies for securing data in the cloud. It ensures that data remains confidential by converting plaintext into an unreadable format, which can only be decrypted by authorized users with the correct key. End-to-end encryption is a critical cloud security feature, as it protects data both in transit and at rest.

Major cloud service providers such as Amazon Web Services (AWS), Microsoft Azure, and Google Cloud offer encryption services to protect sensitive data. AWS, for example, provides encryption mechanisms that cover data at all stages of its lifecycle. Server-Side Encryption (SSE) ensures that data stored in AWS S3 buckets is encrypted automatically using advanced algorithms such as AES-256.[3]

According to Kshetri, "Encryption technologies have become the cornerstone of cloud security, ensuring that unauthorized access to sensitive information is virtually impossible."[4] Additionally, the use of homomorphic encryption, which allows data to be encrypted while still being processed, has emerged as an important advancement in protecting cloud data without compromising performance.

Zero Trust Architecture: A New Approach to Cloud Security

One of the most significant advancements in cloud security is the adoption of Zero Trust Architecture (ZTA). Unlike traditional security models that trust users within the organization's network perimeter, Zero Trust operates on the principle of "never trust, always verify." Every user, device, and application should be continuously authenticated and authorized, regardless of their location within or outside the organization's network.

This shift is crucial in cloud environments, where perimeter-based security models are no longer sufficient due to the distributed nature of cloud resources. According to Forrester Research, "Zero Trust has become essential in protecting cloud environments from both external and internal threats by enforcing strict access control policies at every layer of the cloud infrastructure."[5]

An example of Zero Trust in action is Google's BeyondCorp, a security framework that enables secure access to internal applications without relying on traditional VPNs. Instead, BeyondCorp verifies the identity of users and the health of their devices before granting access, making it ideal for cloud-based environments.[6] This approach ensures that only authenticated users can access cloud resources, significantly reducing the risk of unauthorized access.

Identity and Access Management (IAM): Cloud Security Enhancement

Identity and Access Management (IAM) is another critical component of cloud security, as it ensures that only authorized users have access to cloud resources. IAM systems enable organizations to define and enforce security policies that dictate who can access what data and under what conditions. In cloud environments, where multiple users and devices access data simultaneously, robust IAM practices are crucial for maintaining security.

Cloud providers such as AWS, Azure, and Google Cloud offer comprehensive IAM solutions that allow organizations to manage user permissions, roles, and access policies. Multi-factor authentication (MFA), role-based access control (RBAC), and privileged access management (PAM) are key IAM features that provide an additional layer of protection. MFA, in particular, ensures that even if a user's credentials are compromised, the attacker would still need access to a second authentication factor to gain entry.

According to Alenezi et al., "IAM practices are critical in preventing unauthorized access to cloud resources, particularly as organizations increasingly adopt hybrid and multi-cloud strategies."[7]. In hybrid cloud environments, where data is distributed across on-premises and cloud infrastructures, IAM becomes even more important for securing access points and ensuring that users can only access the resources they are authorized to.

Automated Security Monitoring and Threat Detection

Another advancement in cloud security is the use of automated security monitoring and threat detection systems. Cloud service providers now offer integrated security tools that continuously monitor cloud environments for suspicious activity, misconfigurations, and potential vulnerabilities. These tools use machine learning (ML) and artificial intelligence (AI) to identify and respond to threats in real time. This reduces the time it takes to detect and mitigate attacks.

For example, AWS offers Amazon GuardDuty, an intelligent threat detection service that uses ML algorithms to analyze logs and detect anomalies that may indicate a security breach.[8] Similarly, Microsoft Azure offers Security Center, a unified security management system that provides advanced threat protection across hybrid cloud environments.[9]

These tools allow organizations to respond to threats more quickly and effectively, and minimize the impact of potential security incidents. As AI and ML technologies continue to evolve, cloud security monitoring systems will become even more proactive, enabling automated responses to emerging threats.

Compliance and Regulatory Frameworks for Cloud Security

Cloud security also involves adhering to compliance and regulatory frameworks designed to protect sensitive data. Organizations operating in regulated industries, such as healthcare, finance, and government, should comply with stringent security standards to ensure the protection of personal and financial information. The General Data Protection Regulation (GDPR), Health Insurance Portability and Accountability Act (HIPAA), and Federal Risk and Authorization Management Program (FedRAMP) are some of the key regulations that cloud service providers should comply with.

Cloud providers have developed compliance frameworks to help organizations meet these regulatory requirements. For example, AWS offers a compliance framework that aligns with various global standards, providing detailed guidance on how organizations can secure their cloud environments in accordance with legal requirements.[10] By adhering to these frameworks, organizations can ensure that they are implementing best practices for cloud security while avoiding potential regulatory penalties.

Challenges in Cloud Security

Despite the advancements in cloud security, there are still several challenges that organizations should navigate. One of the most significant challenges is shared responsibility. In cloud environments, security is a shared responsibility between the cloud provider and the customer. While cloud providers are responsible for securing the infrastructure, customers are responsible for securing their applications, data, and access controls.

Additionally, the increasing complexity of multi-cloud and hybrid-cloud environments introduces new risks. Managing security across multiple cloud platforms requires a comprehensive understanding of each provider's security tools and configurations. As Bisong and Rahman explain, "Organizations should implement a unified security strategy that covers all cloud environments to prevent gaps in security."[11]

Conclusion

The advancements in cloud security over the past decade have significantly enhanced the protection of data and applications hosted in the cloud. Technologies such as encryption, Zero Trust architecture, IAM, and automated threat detection have addressed many of the security challenges associated with cloud computing. However, as organizations continue to adopt cloud services, they should remain vigilant and proactive in implementing robust security practices to ensure the confidentiality, integrity, and availability of their cloud-based resources.

2.4 End-to-End Encryption and Secure Communication

End-to-end encryption (E2EE) has become a cornerstone of secure communication in the digital age. E2EE ensures that only the communicating parties can read the messages exchanged between them. E2EE is used in a wide range of applications, from messaging services to online banking. It provides confidentiality and protection against unauthorized access. As cyber threats continue to evolve, E2EE offers a reliable solution for safeguarding sensitive data in transit and protecting it from interception or tampering.

The Fundamentals of End-to-End Encryption

End-to-end encryption works by encrypting data on the sender's device and decrypting it only on the recipient's device. This means that even if a third party intercepts the communication, they will not be able to read the content without the correct decryption key. In traditional encryption methods, data might be encrypted during transit but decrypted at intermediate points, leaving it vulnerable to attacks at those stages. However, with E2EE, the data remains encrypted throughout its entire journey from sender to receiver.

One of the key features of E2EE is its use of public key cryptography, which involves two keys: a public key that is used to encrypt the message and a private key that is used to decrypt it. Only the intended recipient possesses the private key, ensuring that the communication remains secure. According to Schneier, "end-to-end encryption provides a high level of confidentiality, preventing even service providers from accessing the contents of communications."[1]

E2EE in Messaging Applications

The most common use of E2EE is in messaging applications, where users exchange sensitive information, such as personal data, financial details, or private conversations. Popular messaging services like WhatsApp, Signal, and Telegram all implement E2EE to protect user communications from prying eyes. For instance, WhatsApp introduced E2EE in 2016, ensuring that all messages, calls, and shared files are encrypted from the moment they leave the sender's device until they reach the recipient.[2]

E2EE has become a critical feature for messaging platforms because it guarantees that not even the service providers can access the content of the messages. This level of security is particularly important in cases where governments or other entities may request access to user communications. As Pfefferkorn points out, "E2EE makes it impossible for service providers to comply with such requests without undermining the security of their platforms." [3]

E2EE in Financial Transactions and Online Banking

In addition to messaging, E2EE plays a crucial role in securing financial transactions and online banking. Financial institutions rely on E2EE to protect sensitive customer data, such as account numbers, passwords, and transaction details, from being intercepted by cybercriminals. The use of encryption in online banking ensures that even if data is intercepted during transmission, it cannot be decrypted without the proper key.

For example, TLS (Transport Layer Security), a widely adopted encryption protocol, is used to secure communications between customers and banking websites, protecting data from man-in-the-middle (MitM) attacks. TLS ensures that all data transmitted between the client's browser and the bank's servers is encrypted, safeguarding financial information during online transactions.[4]

According to Kshetri, "Encryption protocols like TLS have become indispensable in financial services, where the confidentiality and integrity of transactions are paramount."[5] As financial transactions continue to move online, the demand for robust encryption measures, such as E2EE, will only grow.

E2EE and Secure Communication in Business

Businesses also benefit from E2EE, particularly in securing sensitive corporate communications. With the rise of remote work and the increasing reliance on digital communication tools, organizations should ensure that their communications are protected from interception and espionage. E2EE helps businesses maintain the confidentiality of their internal discussions, intellectual property, and sensitive customer information.

For example, Zoom, a popular video conferencing platform, introduced E2EE in 2020 after facing criticism for security vulnerabilities. By implementing E2EE, Zoom now ensures that video calls, chats, and shared files are protected from unauthorized access, even by the company itself.[6] This shift reflects a growing trend among businesses to adopt E2EE for enhanced security in their communication channels.

According to Liu et al., "E2EE offers businesses a critical advantage in securing communication, especially as cyberattacks targeting corporate espionage become more sophisticated."[7]

As organizations increasingly rely on digital communication, the adoption of E2EE will continue to rise.

Challenges and Limitations of E2EE

Despite its numerous advantages, E2EE is not without challenges. One of the primary concerns is the trade-off between security and usability. While E2EE provides strong protection for data in transit, it can complicate certain features, such as backups and syncing across devices. For example, WhatsApp users can back up their messages to cloud services like iCloud or Google Drive, the backups were not originally encrypted end-to-end, which meant that the cloud provider could access the data. However, WhatsApp has now added the option to enable end-to-end encrypted backups, which means that even when the data is stored in the cloud, users can protect their chat history by requiring a password or encryption key to access it. Unless these backups are encrypted, they remain vulnerable to attacks. This issue highlights the challenge of ensuring full encryption coverage throughout the entire lifecycle of the data.

Another significant challenge is the potential for E2EE to be used by malicious actors to hide their activities. Criminals and terrorists may use encrypted messaging services to plan illicit activities without fear of interception by law enforcement. As a result, some governments have called for backdoor access to E2EE systems, allowing law enforcement agencies to decrypt communications when necessary. However, cybersecurity experts argue that introducing such backdoors would weaken the overall security of the systems, as they could be exploited by attackers.

As Levy and Schneier explain, "Any attempt to introduce backdoors into E2EE systems would create vulnerabilities that could be exploited by malicious actors, undermining the very security these systems are designed to provide."[8] This ongoing debate between privacy advocates and governments continues to shape discussions about the future of encryption.

Future Developments in E2EE Technology

The future of E2EE lies in its continued integration with emerging technologies and the development of more advanced encryption methods. One area of interest is the use of quantum-resistant encryption algorithms to protect data from the potential threats posed by quantum computing. Quantum computers, once fully developed, could break many of the encryption algorithms currently in use, making it essential for researchers to develop new, more secure cryptographic methods.

In response to this threat, researchers are exploring post-quantum cryptography, which aims to create encryption algorithms that can withstand attacks from quantum computers. As Alagic and Fehr point out, "post-quantum cryptography will play a crucial role in ensuring that encryption remains a viable method of protecting data, even in the face of advances in quantum computing."[9]

Additionally, advancements in homomorphic encryption could further enhance E2EE by allowing encrypted data to be processed without being decrypted. This would enable secure data processing in cloud environments, where data needs to be manipulated but should remain confidential throughout the process.

Conclusion

End-to-end encryption has proven to be an indispensable tool for ensuring secure communication in an increasingly digital world. Its applications span from personal messaging services to financial transactions and corporate communications, offering protection against unauthorized access and cyberattacks. Despite the challenges and debates surrounding its use, E2EE remains a critical component of modern cybersecurity, providing a robust defense against the interception of sensitive data. As encryption technologies continue to evolve, E2EE will remain a fundamental pillar in the protection of digital communication.

2.5 Biometrics and Multi-Factor Authentication

The use of biometrics and multi-factor authentication (MFA) in cybersecurity has revolutionized the way individuals and organizations safeguard access to sensitive systems and data. These technologies add layers of security, making it significantly harder for unauthorized users to gain access. As cyber threats grow increasingly sophisticated, biometrics and MFA offer enhanced protection by verifying user identities using factors that go beyond traditional passwords. Their integration into modern cybersecurity strategies reflects the ongoing effort to strengthen authentication mechanisms and reduce the risk of breaches.

Biometrics: Using Unique Human Traits for Authentication

Biometric authentication relies on unique physiological or behavioral characteristics to verify an individual's identity. These traits include fingerprints, facial recognition, iris scans, and voice recognition.

Since these characteristics are difficult to replicate, biometrics offer a higher level of security compared to passwords, which can be easily guessed or stolen. According to Jain et al., "biometrics provide an inherently secure means of authentication because they rely on characteristics that are unique to each individual."[1]

The adoption of biometric technologies has grown significantly in recent years. For instance, Apple's Face ID and Touch ID use facial recognition and fingerprint scanning, respectively, to authenticate users and allow them to unlock their devices, make payments, or access sensitive apps. Similarly, Microsoft Windows Hello provides users with biometric authentication options, including facial recognition and fingerprint scanning, to log in to their devices without needing a password.[2]

Biometric systems are also widely used in financial services, where the need for secure authentication is paramount. Banks and financial institutions have adopted biometric authentication to protect access to customer accounts and secure high-value transactions. For example, HSBC uses voice recognition technology to verify customers' identities during phone banking transactions, ensuring that only authorized individuals can access their accounts.[3]

Multi-Factor Authentication: Combining Multiple Layers of Security

While biometrics alone provide a strong layer of security, multi-factor authentication (MFA) further enhances protection by requiring users to present multiple pieces of evidence to verify their identity. MFA combines two or more authentication factors, such as something the user knows (password), something the user has (smartphone or token), and something the user is (biometrics), to create a more secure authentication process.

MFA is highly effective in mitigating credential-based attacks, such as phishing and brute-force attacks. Even if a user's password is compromised, an attacker would still need access to the second authentication factor, such as a physical token or biometric data, to successfully log in. As Gevers and Jacobs explain, "MFA significantly reduces the likelihood of unauthorized access by requiring multiple, independent authentication factors." [4]

One of the most common forms of MFA is SMS-based authentication, where users receive a one-time password (OTP) via text message to verify their identity. However, due to vulnerabilities in SMS communication, such as SIM swapping and man-in-the-middle attacks, security experts recommend more secure methods, such as time-based one-time passwords (TOTP) generated by authentication apps like Google Authenticator or Microsoft Authenticator. These apps provide a more secure way to deliver OTPs without relying on SMS, which is susceptible to interception. [5]

Biometrics in MFA: A Powerful Combination

Combining biometrics with MFA creates a powerful authentication mechanism that enhances security without sacrificing convenience. For example, a user might be required to scan their fingerprint and enter a password or use facial recognition alongside an OTP to access a secure system. This approach ensures that even if one authentication factor is compromised, the attacker would still need to bypass the biometric layer, which is much more difficult to replicate.

In the financial sector, the use of biometric MFA is becoming more common. MasterCard introduced Identity Check, also known as Selfie Pay, which allows users to authenticate online payments by taking a selfie in addition to entering their password. This biometric MFA solution provides both security and convenience, as it reduces the risk of fraud while making the authentication process more user-friendly.[6]

Similarly, Microsoft Azure Active Directory supports biometric authentication in combination with MFA, enabling organizations to implement secure access controls for their cloud services. This integration allows users to authenticate using a combination of biometric data and other factors, such as passwords or security tokens. This ensures that only authorized individuals can access cloud resources.[7]

Benefits of Biometrics and MFA in Cybersecurity

The primary benefit of biometrics and MFA is the enhanced security they provide. Traditional passwords are inherently insecure, as they can be guessed, stolen, or reused across multiple accounts. By integrating biometrics and MFA, organizations can significantly reduce the risk of credential theft and unauthorized access. According to a study by Verizon, "the use of MFA could prevent up to 99% of credential-based cyberattacks."[8]

Another advantage is the improved user experience. While passwords can be cumbersome to remember and manage, biometrics and MFA streamline the authentication process by allowing users to log in using traits they naturally possess or devices they already carry. This convenience reduces the likelihood of user error, such as writing down passwords or using weak, easily guessable credentials.

Additionally, biometrics and MFA can enhance compliance with regulatory requirements. Many industries, such as healthcare and finance, are subject to strict regulations that mandate strong authentication measures to protect sensitive data. For example, the Health Insurance Portability and Accountability Act (HIPAA) requires healthcare providers to implement secure access controls to safeguard patient information. Biometrics and MFA offer a practical solution for meeting these regulatory requirements while ensuring that data remains protected. [9]

Challenges and Limitations of Biometrics and MFA

Despite their benefits, biometrics and MFA are not without challenges. One concern is the privacy implications of biometric data. Unlike passwords, biometric information is permanent and cannot be changed if compromised. If a biometric database is breached, the consequences can be severe, as attackers could use stolen biometric data to impersonate individuals across multiple systems.

Moreover, while biometrics offer strong authentication, they are not foolproof. Some biometric systems can be spoofed using high-quality images or replicas of fingerprints. In 2019, researchers demonstrated that they could bypass facial recognition systems using 3D-printed masks, highlighting the need for continuous improvements in biometric technologies. [10]

Another challenge is the cost and complexity of implementing biometric systems and MFA. While large organizations may have the resources to deploy these technologies at scale, smaller businesses may struggle with the financial and technical requirements.

Additionally, the integration of MFA into existing systems can be complex, particularly for organizations with legacy infrastructure.

Lastly, individuals who rely on biometric authentication to access their data may unintentionally prevent their successors from accessing critical information, banking systems, or assets in the event of their passing.

The Future of Biometrics and MFA

As cyber threats continue to evolve, biometrics and MFA will play an increasingly important role in the future of cybersecurity. Behavioral biometrics, which analyze patterns such as typing speed, mouse movement, and navigation habits, represent the next frontier in biometric authentication. These technologies can provide continuous authentication by monitoring user behavior in real-time, ensuring that only authorized individuals can access systems, even after initial login.[11]

Furthermore, biometric cryptography offers a promising solution for securing biometric data itself. By combining cryptography with biometric data, it is possible to encrypt biometric templates, ensuring that they cannot be misused or replicated by attackers. As Alzubaidi and Kalita explain, "biometric cryptography adds an extra layer of security by protecting biometric information through encryption, making it even harder for attackers to exploit."[12]

Conclusion

Biometrics and multi-factor authentication have revolutionized the way organizations secure access to sensitive systems and data. By combining multiple layers of security, including unique physical traits and additional authentication factors, these technologies provide robust protection against modern cyber threats. As organizations continue to adopt biometrics and MFA, they will be better equipped to prevent unauthorized access, safeguard sensitive information, and meet regulatory requirements. However, ongoing developments in biometric technology and encryption will be essential to addressing privacy concerns and further enhancing security in the digital age.

2.6 Quantum Cryptography and Its Potential

Quantum cryptography represents a significant leap in the field of cybersecurity, leveraging the principles of quantum mechanics to secure data in ways that are theoretically unbreakable. Unlike classical encryption methods, which rely on complex mathematical problems for security, quantum cryptography uses the laws of physics to protect information. The most well-known application of quantum cryptography is Quantum Key Distribution (QKD), a technique that allows two parties to share a secret key with absolute security. As quantum computing continues to evolve, quantum cryptography offers the potential to transform the landscape of secure communications.

The Basics of Quantum Cryptography

At the heart of quantum cryptography is the concept of quantum superposition, where quantum particles, such as photons, can exist in multiple states simultaneously until they are measured. This principle is crucial for Quantum Key Distribution (QKD), the most widely studied application of quantum cryptography. In QKD, quantum bits (qubits) are transmitted between two parties, typically referred to as Alice and Bob, to establish a shared secret key. Any attempt by an eavesdropper (Eve) to intercept the qubits will disturb their quantum state, making the intrusion detectable.

According to Scarani et al., "quantum cryptography provides a method of key distribution that is fundamentally secure, as any eavesdropping attempt will leave detectable traces."[1] This feature makes QKD immune to the types of attacks that threaten classical encryption methods, such as brute-force attacks or eavesdropping during transmission.

One of the earliest demonstrations of QKD was the BB84 protocol, developed by Charles Bennett and Gilles Brassard in 1984. The BB84 protocol laid the foundation for modern quantum cryptographic systems by showing how quantum mechanics could be used to establish a secure communication channel. Today, QKD systems have advanced significantly, with commercial implementations available to secure critical communications.

Quantum-Safe Encryption: Preparing for the Quantum Threat

The potential of quantum computers to break existing cryptographic systems is one of the key drivers behind the development of quantum cryptography. Classical encryption algorithms, such as RSA and Elliptic Curve Cryptography (ECC), rely on the difficulty of factoring large numbers or solving discrete logarithm problems. However, Shor's algorithm, developed in 1994, demonstrated that a sufficiently powerful quantum computer could solve these problems exponentially faster than classical computers, rendering current encryption methods vulnerable.

As Mosca explains, "The advent of quantum computing threatens to undermine the security of much of today's digital infrastructure, necessitating the development of quantum-safe encryption methods."[2] In response to this threat, researchers are working on post-quantum cryptography, which seeks to develop new encryption algorithms that can withstand quantum attacks. However, while post-quantum cryptography offers a potential solution, it relies on unproven mathematical assumptions, unlike quantum cryptography, which is based on the laws of physics.

Quantum cryptography, particularly QKD, provides a quantum-safe method of key distribution that is immune to attacks from both classical and quantum computers. By securely distributing encryption keys, QKD ensures that even if a quantum computer were to compromise traditional encryption algorithms, the key exchange process would remain secure. This capability is essential for protecting sensitive communications in the post-quantum era.

Applications of Quantum Cryptography in Cybersecurity

Quantum cryptography is poised to play a crucial role in securing critical infrastructure, government communications, and financial transactions. One of the most promising applications of QKD is in securing fiber-optic communications, which are widely used in financial institutions, military operations, and telecommunications. By integrating QKD with existing fiber-optic networks, organizations can ensure that their communication channels are immune to interception, even by quantum adversaries.

For instance, the Swiss government has deployed QKD to secure communications between government offices and military installations, ensuring that sensitive data remains protected against both classical and quantum threats.[3] Similarly, China has established the world's longest QKD network, spanning over 2,000 kilometers between Beijing and

Shanghai, to secure financial transactions and government communications.[4]

In addition to securing fiber-optic networks, quantum cryptography has applications in satellite-based communication. In 2016, China launched Micius, the world's first quantum communications satellite, to test the feasibility of QKD in space. The satellite successfully transmitted quantum keys over a distance of 1,200 kilometers, demonstrating the potential for global quantum-secured communication networks.[5] As quantum cryptography continues to advance, satellite-based QKD could enable secure, global communication networks that are immune to eavesdropping.

Challenges and Limitations of Quantum Cryptography

Despite its promise, quantum cryptography faces several challenges and limitations that should be addressed before it can be widely adopted. One of the primary challenges is the distance limitation of QKD. Quantum signals degrade over long distances, limiting the effective range of QKD systems. While quantum repeaters are being developed to extend the range of QKD networks, these technologies are still in the early stages of development.

Another challenge is the cost and complexity of implementing quantum cryptography. QKD systems require specialized hardware, such as single-photon detectors and quantum light sources, which can be expensive to deploy and maintain. As a result, quantum cryptography is currently more feasible for securing high-value communications, such as those used by governments and financial institutions, rather than everyday consumer applications.

Additionally, quantum cryptography is not a silver bullet for all cybersecurity challenges. While QKD provides a secure method for distributing encryption keys, it does not address other aspects of cybersecurity, such as endpoint security or protecting data at rest. As Bennett and Brassard have pointed out, "Quantum cryptography addresses the specific problem of secure key distribution, but it should be combined with other security measures to provide comprehensive protection."[6]

Future Directions for Quantum Cryptography

As research into quantum cryptography continues, several promising directions are emerging that could overcome its current limitations. One area of focus is the development of quantum repeaters, which would allow QKD signals to be transmitted over much longer distances without degradation. Quantum repeaters work by entangling particles at different points along the communication channel, which enables secure transmission over thousands of kilometers.

Another area of research is the integration of quantum cryptography with Blockchain technology. Blockchain relies on cryptographic algorithms to secure transactions, but these algorithms could be vulnerable to quantum attacks. By incorporating QKD into Blockchain networks, it may be possible to create a quantum-secure Blockchain that is resistant to both classical and quantum adversaries.[7]

In addition, researchers are exploring the potential of quantum-secure authentication methods that could replace traditional passwords and biometrics. These methods would use the principles of quantum mechanics to generate unique, unforgeable authentication tokens that would provide a higher level of security than current systems.

Conclusion

Quantum cryptography offers a revolutionary approach to securing communications, leveraging the laws of quantum mechanics to provide unbreakable encryption. Its ability to detect eavesdropping and protect against quantum computer attacks makes it an essential tool for securing sensitive communications in the post-quantum era. While challenges related to distance limitations, cost, and complexity remain, ongoing research is paving the way for broader adoption of quantum cryptography. As quantum technologies continue to advance, quantum cryptography will play a crucial role in the future of cybersecurity, safeguarding critical infrastructures and global communication networks from both classical and quantum threats.

Test Your Knowledge On Technological Innovations In Cybersecurity

Q1: How is AI used in cybersecurity threat detection?

Q2: What is the role of machine learning in cybersecurity?

Q3: How does Blockchain technology enhance cybersecurity?

Q4: How is Blockchain used to secure data in the IoT ecosystem?

Q5: What are the challenges of securing cloud computing environments?

Q6: What advancements have been made in cloud security?

Q7: How does end-to-end encryption secure communication?

Q8: How does quantum cryptography differ from traditional cryptography?

Q9: What is the potential impact of quantum computing on encryption standards?

Q10: How do biometrics improve authentication systems?

Q11: What role does multi-factor authentication play in improving cybersecurity?

Q12: How does 5G technology introduce new cybersecurity challenges?

Q13: How does edge computing influence cybersecurity strategies?

Q14: What are the security concerns surrounding IoT devices?

Q15: What role does AI play in the automation of cybersecurity defenses?

ANSWERS

Q1: A: AI is used to detect patterns and anomalies in network traffic, identify potential threats in real time, and automate incident responses to reduce the risk of human error.

Q2: A: Machine learning helps improve threat detection systems continuously by learning from previous attacks and adapting defenses accordingly.

Q3: A: Blockchain provides a decentralized and immutable ledger, making it difficult for hackers to alter data or manipulate transaction histories.

Q4: A: Blockchain ensures data integrity and security in IoT devices by providing transparent and tamper-proof records of data exchanges.

Q5: A: Challenges include managing access control, ensuring data privacy, and mitigating the risks of data breaches or misconfigurations in cloud environments.

Q6: A: Cloud security advancements include encryption, multi-factor authentication, zero-trust architectures, and AI-driven anomaly detection.

Q7: A: End-to-end encryption ensures that only the sender and receiver can access the data, protecting it from eavesdroppers during transmission.

Q8: A: Quantum cryptography uses the principles of quantum mechanics to provide ultra-secure encryption that cannot be broken by classical computing methods.

Q9: A: Quantum computing could break traditional encryption methods, necessitating the development of quantum-resistant algorithms to maintain security.

Q10: A: Biometrics, such as fingerprint scanning or facial recognition, offer more secure and personalized authentication methods that are difficult to replicate.

Q11: A: Multi-factor authentication combines something you know (password), something you have (device), and something you are (biometric), significantly reducing unauthorized access.

Q12: A: The speed and complexity of 5G networks expand the attack surface, making it more difficult to monitor and secure all connected devices and data streams.

Q13: A: Edge computing decentralizes data processing, making it critical to secure endpoints where data is processed and stored outside traditional data centers.

Q14: A: IoT devices often lack robust security features, making them vulnerable to attacks that can compromise an entire network or ecosystem.

Q15: A: AI automates threat detection, incident response, and the mitigation of security breaches, reducing response times and improving overall defense strategies.

CHAPTER 3

THE CYBER THREAT LANDSCAPE
Note: All references in this chapter are on page 396 - 398

The Cyber threats have become more complicated and varied over time. This chapter takes a close look at the different types of attacks that have appeared in today's digital world. There are a lot of different types of cyber risks today, from viruses and worms that damage computers to advanced persistent threats (APTs) that get into networks without being seen. We begin by tracing the evolution of malware—malicious software that takes many forms, including Trojans and ransomware. We'll then shift to ransomware attacks, which have recently surged, crippling businesses by encrypting their data and demanding payments. Social engineering techniques like phishing, spear phishing, and vishing target human vulnerabilities rather than technical ones, making them an ever-present danger.

We will also look into the shadowy world of advanced persistent threats (APTs) and nation-state players, who use their skills to spy on or damage computers over a long period of time. On the last part of the chapter, the new risks that come with the Internet of Things (IoT) are talked about. Insider threats are people that an organization trusts but are also unsafe. What kinds of cyber threats do businesses face? By the end of this chapter, you'll have a good idea of these threats and how they use different flaws.

3.1 Evolution of Malware: Viruses, Worms, and Trojans

Malware, short for malicious software, has evolved significantly over the past few decades, becoming one of the most persistent and damaging forms of cyber threats. The most well-known types of malware, viruses, worms, and Trojans have transformed in both complexity and functionality. They evolved from simple disruptive programs to sophisticated tools used for espionage, financial gain, and cyber warfare. Understanding the history and development of these malicious programs is crucial for developing effective defenses in the modern cybersecurity landscape.

The Early Days of Malware: Viruses

The history of malware can be traced back to the early computer viruses of the 1970s and 1980s. A virus is a type of malware that attaches itself to a legitimate program or file in a system and spreads when the infected file is executed. Viruses were initially designed as experiments or pranks, often displaying harmless messages or causing minor disruptions. However, as computers became more integrated into society, viruses grew in both prevalence and severity.

One of the earliest known computer viruses, Creeper, emerged in 1971. It was a self-replicating program. It infected ARPANET (Advanced Research Projects Agency Network), and displayed the message "I'm the creeper: catch me if you can!" This early virus was more of a curiosity than a malicious threat, but it paved the way for more destructive malware in the years to come.[1]

By the late 1980s, viruses had evolved into more harmful programs capable of deleting files, corrupting systems, and causing widespread disruption. The Morris Worm of 1988, while technically a worm (discussed later), demonstrated the potential damage that could be caused by self-replicating programs. The virus landscape continued to evolve with the introduction of macro viruses in the 1990s, which targeted applications like Microsoft Word and spread through infected documents. The Melissa virus in 1999 was one of the most notorious examples, infecting thousands of systems by exploiting Microsoft Word macros and causing millions of dollars in damage.[2]

According to Gordon and Loeb, "the early development of viruses was driven by the challenge of manipulating systems, but as the internet expanded, viruses became powerful tools for financial and political gain "[3]. Today, viruses remain a significant threat, often embedded in email attachments or downloaded from compromised websites, where they can still wreak havoc on unsuspecting users.

Worms: Autonomous and Fast-Spreading Threats

Unlike viruses, which require human intervention to spread, worms are self-replicating programs that can propagate independently. Worms exploit vulnerabilities in networked systems, allowing them to move from one device to another without any user action. This ability to spread autonomously makes worms particularly dangerous, as they can infect vast numbers of systems in a short period.

One of the earliest and most infamous examples of a worm is the Morris Worm, created by Robert Tappan Morris in 1988. The worm was intended to gauge the size of the early internet but ended up infecting around 10% of the computers connected to ARPANET, causing widespread disruption.[4] The Morris Worm exploited vulnerabilities in UNIX-based systems, demonstrating how worms could cause significant damage by targeting network weaknesses.

In the years since the Morris Worm, worms have evolved to become more sophisticated and destructive. One of the most devastating examples is the ILOVEYOU worm of 2000, which spread via email and caused over $10 billion in damage worldwide. The worm infected millions of computers by exploiting a vulnerability in Microsoft Outlook, demonstrating the ability of worms to spread quickly and cause massive disruption.[5]

More recently, worms have been used in cyber warfare and espionage. The Stuxnet worm, discovered in 2010, is one of the most sophisticated examples of a worm used for state-sponsored attacks. Stuxnet targeted Iran's nuclear facilities, causing physical damage to centrifuges by exploiting vulnerabilities in Siemens' industrial control systems. The worm spread through USB drives and infected thousands of systems before its true purpose was discovered [6]. According to Rid and McBurney, "Stuxnet marked a new era in malware development, where worms are not just tools of disruption but weapons of geopolitical significance."[7]

Trojans: The Hidden Threat

While viruses and worms are designed to spread and disrupt systems, Trojans are more insidious, often disguising themselves as legitimate software to trick users into installing them. Once installed, a Trojan can open backdoors for attackers, steal sensitive information, or grant unauthorized access to the system. Trojans do not replicate themselves like viruses or worms but rely on deception and social engineering to spread.

The name "Trojan" comes from the Trojan Horse of Greek mythology, reflecting the way these programs hide their malicious intent behind a seemingly benign exterior. Trojans often masquerade as useful software, such as antivirus programs or system updates, convincing users to install them willingly. Once installed, they can steal passwords, record keystrokes, or even remotely control the victim's device.

One of the earliest examples of a Trojan is the AIDS Trojan (also known as the PC Cyborg Trojan) from 1989. This Trojan was distributed on floppy disks under the guise of an AIDS information program. Once installed, the Trojan encrypted files on the victim's computer and demanded payment to restore access, making it one of the first known examples of ransomware.[8]

Trojans have since evolved to become more sophisticated and difficult to detect. Zeus is a well-known Trojan that has been used in large-scale cybercrime operations. First discovered in 2007, Zeus primarily targeted banking systems, stealing millions of dollars by capturing login credentials for online banking accounts.[9] Modern Trojans are often part of botnets, networks of infected computers that can be controlled remotely by cybercriminals for tasks such as launching Distributed Denial-of-Service (DDoS) attacks or spreading spam.

According to Choo, "Trojans are increasingly used in combination with other types of malware, creating complex, multi-stage attacks that are difficult to detect and mitigate."[10] The rise of fileless malware, which operates entirely in a system's memory without leaving a trace on the hard drive, has made Trojans even harder to identify using traditional antivirus tools.

The Evolution of Malware: From Nuisance to Weapon

The evolution of viruses, worms, and Trojans reflects the broader shift in the motivations behind malware development. In the early days, malware was often created by hobbyists or hackers seeking to demonstrate their technical prowess. However, as the internet has become more integral to daily life and business, malware has evolved into a tool for financial gain, espionage, and even warfare.

Today's malware is more sophisticated than ever, often combining elements of viruses, worms, and Trojans to create multi-vector attacks that target multiple aspects of a system's defenses. These attacks are often carried out by organized crime groups, nation-states, or advanced persistent threat (APT) actors, who use malware to steal intellectual property, disrupt critical infrastructure, or gain a competitive advantage.

As Schneier explains, "The evolution of malware reflects the increasing sophistication of cybercriminals and the growing importance of digital assets, making malware one of the most significant threats to modern cybersecurity "[11]. It is important now than ever for cybersecurity professionals to develop more advanced tools and strategies to defend against these increasingly complex threats.

Conclusion

The evolution of malware such as viruses, worms, and Trojans illustrates the ever-changing landscape of cyber threats. From the early viruses that spread through floppy disks to the sophisticated worms and Trojans used in modern cyber warfare, malware has become a powerful tool for cybercriminals and nation-states alike. It is of paramount necessity that the strategies used to detect and defend against malware threats. Understanding the history and development of malware is critical for anticipating future threats and protecting against the growing array of cyberattacks.

3.2 Ransomware Attacks and Their Economic Impact

Ransomware has emerged as one of the most damaging and costly forms of cyberattacks in the modern cyber threat landscape. By encrypting victims' data and demanding payment in exchange for a decryption key, ransomware not only disrupts operations but also imposes significant financial burdens on individuals, businesses, and governments. The increasing frequency and sophistication of ransomware attacks have made them a top concern for cybersecurity professionals, and their economic impact continues to escalate each year.

The Rise of Ransomware Attacks

Ransomware first gained prominence in the early 2000s, but its prevalence skyrocketed with the rise of cryptocurrencies like Bitcoin, which allow attackers to demand payments anonymously. The AIDS Trojan in 1989 is often cited as the first known ransomware attack, although it lacked the sophistication and widespread impact of modern ransomware. In recent years, ransomware has evolved into a lucrative business model for cybercriminals, with attacks becoming more targeted and financially motivated.

A pivotal moment in the evolution of ransomware came with theWannaCry attack in May 2017. This global ransomware outbreak infected over 200,000 computers in more than 150 countries, causing widespread disruption. The attackers exploited a vulnerability in Microsoft Windows systems, known as EternalBlue, to rapidly propagate the malware across networks. The economic impact of WannaCry was substantial, with losses estimated at over $4 billion, affecting industries ranging from healthcare to transportation.[1] According to Greenberg, "WannaCry demonstrated the devastating potential of ransomware to cripple organizations on a global scale, highlighting the urgent need for robust cybersecurity defenses."[2]

The Financial Burden of Ransom Payments

One of the most immediate economic consequences of ransomware attacks is the payment of ransoms. Attackers typically demand payment in cryptocurrencies like Bitcoin, which provide anonymity and make transactions difficult to trace. Ransom demands can range from a few hundred dollars for attacks targeting individuals to millions of dollars in cases involving large corporations or government entities.

In 2021, the Colonial Pipeline ransomware attack became one of the most high-profile cases, in which the company paid a ransom of $4.4 million to the attackers to restore access to its systems. The attack caused a major disruption to fuel supplies along the eastern United States, leading to economic losses far exceeding the ransom itself.[3] The U.S. Department of Justice later recovered a portion of the ransom, but the incident highlighted the broader economic vulnerabilities posed by ransomware attacks on critical infrastructure.

Paying the ransom, however, does not guarantee a full recovery. In many cases, attackers fail to provide functional decryption keys or they demand additional payments before providing the functional decryption keys. According to a report by Sophos, "over 40% of organizations that pay the ransom are unable to recover all their data."[4] This creates further economic strain on victims, who should also invest in recovery efforts and additional cybersecurity measures to prevent future attacks.

Downtime and Operational Disruption

Beyond ransom payments, one of the most significant costs associated with ransomware attacks is the downtime and operational disruption that organizations experience. When critical systems are taken offline by ransomware, businesses can lose access to vital data and services, leading to halted production, missed revenue opportunities, and damaged customer trust.

In the healthcare sector, ransomware attacks can have life-threatening consequences. The WannaCry attack hit the National Health Service (NHS) in the U.K. particularly hard, forcing hospitals to cancel appointments and surgeries and diverting emergency patients to other facilities. The total cost of the attack to the NHS was estimated at £92 million, highlighting how ransomware can disrupt essential services.[5] According to Van Eeten, "the economic cost of downtime is often the most significant and overlooked aspect of ransomware attacks, particularly for organizations in critical industries such as healthcare, energy, and finance."[6]

The manufacturing sector is another frequent target of ransomware attacks due to its reliance on industrial control systems (ICS). In 2019, Norsk Hydro, a major aluminum producer, suffered a ransomware attack that disrupted its operations globally, resulting in losses estimated at $60 million [7]. The company's quick decision not to pay the ransom, coupled with strong incident response measures, allowed it to recover without further financial damage, but the case underscored the vulnerability of manufacturing operations to ransomware.

Indirect Costs and Long-Term Impacts

Ransomware attacks often incur indirect costs that extend far beyond the immediate ransom payment or downtime. Organizations should invest in recovery efforts, including forensic investigations, system restoration, and network security enhancements. These efforts can be time-consuming and costly, especially for small and medium-sized enterprises (SMEs) that lack the resources to recover quickly.

In addition, ransomware attacks can result in reputational damage, particularly if customer data is compromised or service disruptions are prolonged. Customers and business partners may lose trust in a company's ability to safeguard sensitive information, leading to long-term financial consequences. For instance, after a ransomware attack on CWT Global, a travel management company, it was revealed that the personal data of employees and clients had been compromised, raising concerns about data privacy and security.[8]

Furthermore, many organizations face regulatory penalties following ransomware attacks, especially if they fail to adequately protect personal data in accordance with privacy laws like the General Data Protection Regulation (GDPR) or the Health Insurance Portability and Accountability Act (HIPAA). Fines for non-compliance can be substantial, adding to the financial burden of recovering from an attack.

Ransomware-as-a-Service (RaaS): A Lucrative Business Model

The rise of Ransomware-as-a-Service (RaaS) has lowered the barrier to entry for cybercriminals, contributing to the rapid increase in ransomware attacks. RaaS platforms allow criminals with limited technical expertise to purchase or lease ransomware tools from skilled developers in exchange for a share of the ransom profits. This model has made ransomware attacks more accessible and widespread, driving up the overall economic impact of ransomware globally.

According to Kharraz and Robertson, "the commercialization of ransomware through RaaS platforms has democratized cybercrime, leading to a proliferation of attacks on businesses, governments, and individuals alike."[9] RaaS operators continue to innovate, developing more sophisticated malware variants that can evade detection and encrypt files more effectively. This trend has significantly increased the economic toll of ransomware attacks, with global losses estimated at over $20 billion in 2021 alone.[10]

The Broader Economic Impact of Ransomware

The economic impact of ransomware extends beyond the immediate victims, as attacks can have ripple effects throughout the broader economy. For example, the Colonial Pipeline attack disrupted fuel supplies across much of the eastern U.S., causing gas shortages and price increases that affected businesses and consumers alike. Similarly, ransomware attacks on the education sector have forced schools and universities to close temporarily, disrupting students' education and imposing additional costs for system restoration.[11]

In some cases, ransomware attacks can also impact national security. Governments and defense organizations that rely on critical infrastructure are increasingly concerned about the potential for ransomware to disrupt essential services. The U.S. Department of Homeland Security has identified ransomware as a top cyber threat, warning that attacks on critical infrastructure could have severe economic and security implications.[12]

Preventing and Mitigating the Impact of Ransomware

To mitigate the economic impact of ransomware, organizations should adopt a multi-layered cybersecurity strategy that includes robust backup and recovery systems, employee training on phishing attacks, and proactive network monitoring. Cyber insurance is also becoming an important tool for organizations to offset the financial risks associated with ransomware attacks. However, as ransomware attacks increase in frequency and severity, some insurers are raising premiums or limiting coverage for ransomware-related incidents.

According to Kovacs, "cyber insurance can provide organizations with a safety net in the event of a ransomware attack, but it should not be seen as a substitute for strong cybersecurity measures."[13] Organizations should focus on prevention and resilience to minimize the likelihood of falling victim to ransomware and ensure they can recover quickly if an attack occurs.

Conclusion

Ransomware has evolved from a nuisance into one of the most dangerous and costly forms of cybercrime. The economic impact of ransomware is far-reaching, encompassing ransom payments, downtime, indirect costs, and long-term damage to reputation and customer trust. As ransomware attacks become more frequent and sophisticated, organizations should invest in comprehensive cybersecurity strategies to protect themselves from the financial and operational risks posed by this growing threat. Failure to do so could result in devastating economic consequences, both for individual, organizations and the broader economy.

3.3 Social Engineering and Phishing Techniques

Social engineering, particularly **phishing**, has become one of the most prevalent and dangerous forms of cyberattack in recent years. By manipulating human behavior and exploiting psychological vulnerabilities, cybercriminals use social engineering to deceive individuals into divulging sensitive information or performing actions that compromise security. Phishing, a subset of social engineering, primarily involves tricking users into revealing passwords, financial information, or other sensitive data through fraudulent emails, websites, or messages. Understanding the various techniques used in social engineering and phishing is crucial for defending against these attacks and protecting digital assets.

The Fundamentals of Social Engineering

At its core, social engineering is the psychological manipulation of individuals into performing actions or divulging confidential information. Unlike traditional hacking methods, which focus on exploiting technical vulnerabilities, social engineering targets human weaknesses. Attackers rely on trust, fear, curiosity, or urgency to manipulate their victims. According to Mitnick and Simon, "the weakest link in any security system is often the human element, as people are inherently prone to making mistakes or succumbing to manipulation."[1]

Social engineering attacks often take advantage of authority bias, where individuals comply with perceived authority figures, or scarcity, where the fear of missing out leads to rash decisions. For instance, an attacker posing as an IT support staff member might convince a victim to provide their login credentials by claiming that immediate action is needed to prevent a security breach.

One of the most famous examples of social engineering is the attack on RSA Security in 2011, where employees were tricked into opening malicious email attachments. This resulted in the compromise of RSA's SecureID authentication tokens, impacting numerous organizations worldwide.[2] The RSA attack illustrates how even well-trained professionals can fall victim to sophisticated social engineering schemes.

Phishing: The Most Common Social Engineering Attack

Phishing is the most common and widespread form of social engineering. In a phishing attack, cybercriminals masquerade as legitimate entities to deceive individuals into providing personal information, such as usernames, passwords, or credit card numbers. Phishing attacks typically involve emails or messages that appear to come from reputable organizations, such as banks, government agencies, or online services.

One of the earliest and most notable phishing attacks occurred in the early 2000s when attackers sent emails purporting to be from PayPal. The emails asked recipients to verify their accounts by clicking a link and entering their login credentials on a fake website. Many users fell victim to this scam, leading to widespread financial losses and the emergence of phishing as a mainstream attack method [3].

Spear Phishing: Targeted Attacks

Spear phishing is a more targeted version of phishing, where attackers focus on a specific individual or organization. Spear phishing emails are often tailored to the victim, containing personalized information that makes the attack more convincing. Cybercriminals conducting spear phishing attacks often gather information about their targets from social media, company websites, or other publicly available sources.

One high-profile spear phishing attack targeted John Podesta, the chairman of Hillary Clinton's 2016 presidential campaign. The attackers sent a fraudulent email posing as a security alert from Google, tricking Podesta into providing his email password. This led to the leak of thousands of sensitive campaign emails, causing significant political fallout.[4] Spear phishing attacks like this one demonstrate the potentially devastating consequences of highly targeted social engineering schemes.

According to Jagatic et al., "spear phishing is one of the most effective forms of social engineering because it leverages personal information to gain the trust of the victim, making detection much more difficult."[5] The sophistication nature of spear phishing attacks has increased in recent years, with cybercriminals using deepfakes, artificial intelligence, and machine learning to craft convincing messages and even mimic the voices of trusted contacts.

Whaling: Targeting High-Profile Individuals

A specific form of spear phishing, known as whaling, targets high-profile individuals such as CEOs, CFOs, or government officials. Whaling attacks often involve emails that appear to come from trusted sources within the organization and are designed to convince the victim to authorize large financial transfers or provide sensitive business information.

For example, in 2016, Ubiquiti Networks fell victim to a whaling attack that cost the company $46 million. The attackers impersonated company executives and requested fraudulent wire transfers to offshore accounts. Ubiquiti later recovered some of the funds, but the incident highlighted the effectiveness of whaling in targeting senior executives [6]. According to Parsons et al., "whaling attacks are particularly dangerous because they exploit the authority and trust placed in high-level executives, making it easier for attackers to manipulate them into taking damaging actions."[7]

Vishing and Smishing: Expanding Phishing Beyond Email

While phishing traditionally occurs through email, attackers have expanded their techniques to other communication channels, leading to the rise of vishing (voice phishing) and smishing (SMS phishing). Vishing involves phone calls where attackers pose as legitimate entities, such as banks or government agencies, to extract personal information from victims. Smishing uses SMS text messages to deliver phishing attempts, often including a malicious link that directs victims to a fake website.

In one prominent vishing attack in 2020, attackers used social engineering to trick employees at Twitter into providing their login credentials over the phone. The attackers then gained access to high-profile Twitter accounts, including those of Barack Obama, Elon Musk, and Bill Gates, and used these accounts to promote a cryptocurrency scam.[8]

Vishing and smishing attacks are particularly concerning because individuals are often less suspicious of phone calls and text messages compared to emails. According to Hong, "As phishing attacks move beyond email, it becomes increasingly difficult for individuals to recognize fraudulent attempts, especially when attackers use multiple communication channels to create a sense of urgency or trust."[9]

Business Email Compromise (BEC): A Growing Threat

Business Email Compromise (BEC) is a sophisticated form of phishing that targets businesses to trick employees into transferring funds or revealing sensitive information. BEC attacks typically involve impersonating an executive or business partner through a compromised or spoofed email account. These attacks can be financially devastating, as they often involve large sums of money and go undetected until it is too late.

The FBI reported that BEC scams caused $1.8 billion in losses in 2020, making it one of the most financially damaging forms of cybercrime.[10] One of the largest BEC incidents occurred in 2019 when Toyota Boshoku Corporation, a supplier for Toyota, lost $37 million after falling victim to a BEC scam that impersonated an executive and requested a fraudulent wire transfer.[11]

According to Hutchings and Clayton, "BEC attacks are highly effective because they exploit trust within organizations, often bypassing traditional security measures by targeting human weaknesses rather than technological vulnerabilities."[12]

Preventing Social Engineering and Phishing Attacks

Defending against social engineering and phishing requires a combination of technology and awareness training. Technical solutions such as email filtering, multi-factor authentication, and intrusion detection systems can reduce the likelihood of phishing attacks reaching users. Additionally, educating employees about the tactics used by cybercriminals is equally important.

Phishing awareness training programs teach individuals how to recognize phishing attempts, avoid clicking on suspicious links, and verify the legitimacy of communications. According to a Verizon Data Breach Investigations Report, "social engineering attacks are responsible for 22% of all breaches, but effective training can significantly reduce the risk by improving employees' ability to identify and report suspicious activities."[13]

Conclusion

Social engineering and phishing techniques continue to evolve and becoming more sophisticated and difficult to detect. By exploiting human psychology, cybercriminals can bypass even the most advanced technical defenses, making social engineering one of the most effective methods for breaching security systems. As attackers expand their tactics beyond traditional phishing emails to include spear phishing, whaling, vishing, and smishing, organizations should prioritize both technical defenses and human-focused training to mitigate the risks posed by these pervasive threats.

3.4 Advanced Persistent Threats (APTs) and Nation-State Actors

Advanced Persistent Threats (APTs) represent one of the most sophisticated and damaging forms of cyberattacks, typically orchestrated by nation-state actors. Unlike traditional cyberattacks, which may aim for immediate financial gain or disruption, APTs are characterized by their stealth, persistence, and long-term objectives. These attacks often target governments, critical infrastructure, and large corporations, aiming to steal sensitive data, conduct espionage, or sabotage systems. APTs have become an essential tool for nation-states seeking to exert influence, gather intelligence, or weaken their adversaries through cyberspace.

Understanding Advanced Persistent Threats

APTs differ from other forms of cyberattacks in their strategic objectives and operational tactics. Unlike a malware attack, which may be detected and resolved quickly, APTs are designed to remain undetected for extended periods. Once inside a network, attackers establish a foothold and exfiltrate data or manipulate the systems without drawing attention. The persistent nature of these threats allows attackers to gather intelligence, steal intellectual property, or even prepare for future sabotage.

According to Rid and Buchanan, "APTs are distinct from conventional cyberattacks due to their sophisticated techniques, long-term objectives, and often state-sponsored nature."[1] Nation-state actors use APTs for cyber espionage, intellectual property theft, and cyber warfare, giving them strategic advantages in political, military, and economic domains.

One of the most notable characteristics of APTs is the level of customization involved. Attackers often tailor their tools and methods to specific targets, employing zero-day vulnerabilities and custom malware to avoid detection. APT groups frequently use phishing, social engineering, and watering hole attacks to gain initial access to their targets. These tactics involve exploiting trust within an organization, either by targeting individuals through fake emails or by compromising legitimate websites that the target frequents.

APT Groups and Nation-State Actors

Several APT groups have been identified as being affiliated with specific nation-states, often carrying out attacks on behalf of their governments. These groups are typically well-funded and have access to advanced tools and resources, allowing them to execute complex and sustained campaigns.

One of the most infamous APT groups is APT28, also known as Fancy Bear, which is widely believed to be linked to Russian military intelligence (GRU). APT28 has been involved in several high-profile attacks, including the 2016 Democratic National Committee (DNC) breach, where sensitive emails were stolen and leaked during the U.S. presidential election. The group has also targeted government agencies, military organizations, and media outlets in Europe and the United States.[2]

Similarly, APT29, also known as Cozy Bear, is another Russian-linked group believed to be associated with the Federal Security Service (FSB). APT29 has been linked to cyber espionage campaigns targeting the U.S. government, healthcare institutions, and research organizations.

In 2020, APT29 was implicated in attempts to steal COVID-19 vaccine research from U.S., Canadian, and U.K. organizations.[3]

China is another key player in the APT landscape, with several well-known APT groups linked to Chinese intelligence services. APT10, also known as Stone Panda, has been involved in espionage campaigns targeting intellectual property and sensitive commercial data. The group has been particularly active in targeting industries such as aviation, biotechnology, and telecommunications. In 2018, APT10 was accused of stealing trade secrets from companies in over a dozen countries as part of a global espionage campaign. [4]

The Tactics, Techniques, and Procedures (TTPs) of APTs

APTs are known for their advanced Tactics, Techniques, and Procedures (TTPs), which allow them to remain undetected while achieving their objectives. One of the key tactics used by APT groups is spear-phishing, where attackers craft highly targeted emails to trick specific individuals within an organization into providing login credentials or downloading malicious software. Once initial access is gained, attackers often use privilege escalation techniques to obtain higher-level permissions and move laterally across the network.

According to Zander et al., "APTs rely on a combination of stealth, persistence, and technical sophistication, using techniques such as privilege escalation, lateral movement, and data exfiltration to achieve their goals."[5] These techniques make APTs difficult to detect, as attackers often mimic legitimate user behavior or exploit weaknesses in the organization's security posture.

APTs also frequently employ watering hole attacks, where attackers compromise legitimate websites that are frequently visited by their targets. By infecting these sites with malware, attackers can gain access to the systems of anyone who visits the compromised site. This tactic was used in the Operation Aurora attack in 2009, which targeted Google and other major corporations. Attackers compromised a website used by employees of these companies, allowing them to infiltrate their systems and steal intellectual property. [6]

The Role of APTs in Cyber Espionage and Cyber Warfare

APTs are often used for cyber espionage, where attackers seek to gather sensitive information from government agencies, research institutions, or corporations. The stolen data can be used to advance a nation's political, economic, or military objectives. For example, Chinese APT groups have been linked to campaigns targeting intellectual property in industries such as aerospace, pharmaceuticals, and energy, with the goal of advancing China's economic and technological development.[7]

In addition to espionage, APTs are increasingly used as tools of cyber warfare. Nation-state actors may use APTs to disrupt critical infrastructure, weaken an adversary's military capabilities, or sow discord in foreign governments. For instance, the Stuxnet worm, which is widely believed to have been developed by the U.S. and Israeli governments, targeted Iran's nuclear enrichment facilities. The worm sabotaged centrifuges used in the enrichment process, setting back Iran's nuclear program and demonstrating how APTs can be used to cause physical damage through cyberattacks. [8]

According to Clarke and Knake, "the use of APTs in cyber warfare represents a new frontier in conflict, where nation-states can achieve strategic objectives without resorting to conventional military force." [9] The covert nature of APTs allows nation-states to engage in hostilities without directly attributing the attack, complicating international relations and deterrence strategies.

Defending Against APTs

Defending against APTs requires a multi-layered cybersecurity strategy that combines technical defenses with threat intelligence and employee training. Organizations should implement network segmentation, intrusion detection systems (IDS), and privileged access management (PAM) to limit the ability of attackers to move laterally within a network. Regular patching and vulnerability management are also critical in preventing APT groups from exploiting known security weaknesses.

Threat intelligence is essential for identifying and mitigating APT attacks. By analyzing the TTPs of known APT groups, organizations can develop proactive defenses and monitor for indicators of compromise (IOCs) associated with specific threat actors. For example, understanding the spear-phishing techniques used by APT28 or the lateral movement tactics employed by APT29 can help organizations detect and block these attacks before they succeed.

Additionally, employee training is vital in defending against social engineering attacks that APTs frequently use to gain initial access. Teaching employees to recognize phishing emails, suspicious downloads, and other common attack vectors can significantly reduce the risk of compromise. According to a report by FireEye, "human error remains one of the primary vulnerabilities exploited by APTs, making security awareness training a critical component of any defense strategy."[10]

Conclusion

Advanced Persistent Threats (APTs) represent one of the most significant challenges in modern cybersecurity as it`s driven by the strategic objectives of nation-state actors. The global nature of APTs, combined with their potential for long-term impact, makes them a critical concern for governments, businesses, and cybersecurity professionals alike. These sophisticated attacks are not only designed to steal sensitive data but also to undermine the economic, political, and military power of adversaries. As APT groups continue to evolve and adapt their tactics, organizations should develop comprehensive defense strategies that incorporate both technical solutions and human-focused training.

3.5 Insider Threats and Privileged Access Management

Insider threats represent one of the most challenging security risks to mitigate, as they involve individuals with authorized access to an organization's critical systems and sensitive data. Unlike external attacks, insider threats originate from within the organization, making them difficult to detect and prevent. These threats can be malicious, where an employee or contractor intentionally causes harm, or unintentional, where users inadvertently compromise security through negligence or mistakes. Coupled with insider threats, privileged access management (PAM) has become a crucial strategy in protecting organizations from the risks associated with users who have elevated access to sensitive systems.

The Nature of Insider Threats

Insider threats can stem from various sources, including current or former employees, contractors, business partners, or anyone with authorized access to the organization's network. These individuals may misuse their access intentionally, seeking personal financial gain, revenge, or espionage, or they may unintentionally expose the organization to cyber risks through negligence. As noted by Greitzer et al., "insider threats are particularly dangerous because they come from individuals who have already bypassed many of the organization's security measures."[1]

A well-known example of an insider threat occurred in 2013 when Edward Snowden, a former contractor for the U.S. National Security Agency (NSA), leaked classified information detailing the extent of U.S. government surveillance programs. Snowden had access to privileged systems and that enabled him to collect and disclose a vast amount of sensitive data without detection for an extended period. [2] This case illustrates the potential damage that can be caused by insiders with elevated access.

Types of Insider Threats

Insider threats can be classified into three main categories: malicious insiders, negligent insiders, and compromised insiders.

- Malicious insiders are individuals who deliberately abuse their access to cause harm to the organization. These actions may include stealing intellectual property, sabotaging systems, or sharing confidential data with competitors or foreign governments. According to Cappelli et al., "malicious insiders pose a significant risk because they often have both the knowledge and access needed to inflict serious damage."[3]

- Negligent insiders are employees or contractors who unintentionally expose the organization to risks by failing to follow security protocols, misplacing sensitive data, or falling for phishing schemes. While their actions are not malicious, the consequences can be just as severe. A 2020 Ponemon Institute study found that negligent insiders were responsible for over 60% of insider incidents, primarily due to human error or lack of awareness.[4]

- Compromised insiders are individuals whose accounts or credentials have been hijacked by external attackers. In this case, the insider may not even be aware that their access is being used for malicious purposes. Credential theft through phishing or malware can enable attackers to use privileged accounts to move laterally within a network, bypassing security defenses.

The Role of Privileged Access in Insider Threats

Privileged accounts represent one of the most significant vulnerabilities when it comes to insider threats. These accounts provide elevated access to critical systems, databases, and infrastructure, enabling administrators and users with specific roles to perform sensitive tasks such as managing network configurations, modifying system settings, or accessing confidential data. If misused or compromised, privileged accounts can lead to devastating breaches.

According to Gartner, "Privileged access accounts are prime targets for both malicious insiders and external attackers because they hold the keys to the most critical parts of the network."[5] For this reason, privileged access should be carefully managed and monitored to reduce the risk of insider abuse or compromise.

In the Target data breach of 2013, attackers gained access to the company's network by compromising the credentials of an HVAC vendor. Once inside the network, the attackers escalated their privileges and installed malware on point-of-sale systems, eventually stealing 40 million credit card numbers.[6] This breach highlights the risks associated with excessive or poorly monitored privileged access.

Privileged Access Management (PAM): A Key Defense Mechanism

To combat the risks associated with insider threats, organizations should implement Privileged Access Management (PAM) systems. PAM solutions are designed to control, monitor, and audit the use of privileged accounts, ensuring that elevated access is only granted when necessary and is subject to strict oversight. By limiting the number of users with privileged access and enforcing policies such as least privilege, where users only have the minimal level of access required to perform their duties, organizations can significantly reduce the risk of insider threats.

PAM solutions typically include features such as credential vaulting, which stores privileged account credentials in a secure, encrypted location, and session monitoring, which allows administrators to track and record privileged sessions in real time. If suspicious activity is detected, the session can be terminated or flagged for further investigation. As noted by Bertino and Sandhu, "PAM is essential for organizations seeking to safeguard their most critical assets by ensuring that privileged access is tightly controlled and continuously monitored."[7]

Furthermore, PAM systems can enforce multi-factor authentication (MFA) for privileged accounts, requiring users to provide additional verification before gaining access to sensitive systems. This adds an extra layer of security, making it more difficult for compromised credentials to be exploited by external attackers or malicious insiders.

Mitigating Insider Threats Through Policy and Technology

In addition to implementing PAM solutions, organizations should adopt a holistic approach to mitigating insider threats. This includes developing clear security policies, conducting background checks for employees with access to sensitive data, and providing ongoing security awareness training to reduce negligence.

Regular auditing and monitoring of privileged accounts are also essential. Automated tools can help detect unusual activity, such as unauthorized access attempts or the use of privileged accounts outside of normal working hours. Organizations should also implement role-based access control (RBAC) to limit access to sensitive data based on an employee's specific role and responsibilities. This ensures that only those who need access to critical systems can obtain it.

One example of the successful implementation of PAM is Centrify, a cybersecurity company that specializes in identity and access management. Centrify's PAM solution includes features such as just-in-time access, which grants privileged users access only when necessary and automatically revokes access once the task is completed. This approach ensures that privileged access is tightly controlled, reducing the risk of both malicious and negligent insiders.[8]

The Future of Insider Threats and PAM

As organizations continue to adopt cloud services, remote work, and third-party integrations, the attack surface for insider threats expands.

Insiders now have access to an increasingly diverse array of systems, applications, and data, which presents new challenges for cybersecurity professionals. The shift toward remote work has also blurred the boundaries between personal and professional devices, making it harder for organizations to enforce strict security controls.

To address these challenges, the future of PAM will likely involve greater integration with artificial intelligence (AI) and machine learning (ML) to detect insider threats more effectively. AI-powered systems can analyze patterns of behavior across privileged accounts and identify anomalies that could indicate malicious or negligent activity. These systems will also be capable of predictive analytics, enabling organizations to detect potential insider threats before they result in data breaches or other security incidents.

According to Gheyas and Abdallah, "AI-driven PAM solutions will play a crucial role in protecting organizations from insider threats by enabling real-time monitoring and automatic responses to suspicious activity."[9] The adoption of AI and ML technologies in PAM is expected to revolutionize how organizations manage privileged access, providing enhanced security and reducing the risks associated with insider threats.

Conclusion

Insider threats remain one of the most significant challenges in cybersecurity, particularly due to the difficulty in detecting and mitigating attacks that originate from within an organization. By implementing Privileged Access Management (PAM) systems, organizations can limit the potential damage caused by both malicious and negligent insiders. PAM solutions provide critical controls, such as least privilege, session monitoring, and multi-factor authentication that help protect privileged accounts and sensitive data from misuse. It critical that organizations adopt both policy-driven and technology-based approaches to safeguard their most valuable assets.

3.6 Emerging Threats in the IoT and Mobile Landscapes

As the Internet of Things (IoT) and mobile devices become increasingly integrated into daily life and business operations, they also present significant new challenges for cybersecurity. These technologies bring unprecedented convenience and connectivity, but they also expand the attack surface available to cybercriminals. The proliferation of IoT devices and the dependence on mobile technologies have introduced a range of emerging threats, from device hijacking to data breaches, which require innovative security solutions.

The Internet of Things: A Growing Attack Surface

The Internet of Things refers to the network of interconnected devices that communicate and share data over the internet. These devices include everything from smart home gadgets like thermostats and cameras to industrial systems such as manufacturing sensors and healthcare monitors. By 2025, it is estimated that there will be over 75 billion IoT devices worldwide.[1] This growth is largely driven by the demand for automation, efficiency, and enhanced data collection in various sectors.

However, this explosive growth also introduces security vulnerabilities. IoT devices are often designed with convenience in mind and with security as an afterthought. Many devices are shipped with default passwords that users do not change, leaving them vulnerable to attack. Furthermore, IoT devices are often deployed in large numbers, making it difficult to maintain proper security configurations for all devices in a network. As highlighted by Yang et al., "the increasing ubiquity of IoT devices presents a significant attack surface, with each connected device acting as a potential entry point for cybercriminals."[2]

A notable example of the security risks posed by IoT devices is the Mirai botnet attack in 2016. Mirai targeted IoT devices such as security cameras and routers, exploiting weak default passwords to take control of the devices. Once compromised, these devices were used to launch massive Distributed Denial-of-Service (DDoS) attacks, temporarily disrupting internet services across the globe.[3] The Mirai botnet attack demonstrated how vulnerable IoT devices can be if not properly secured.

Device Hijacking and Remote Control

One of the most concerning threats in the IoT landscape is device hijacking, where attackers gain unauthorized control over an IoT device. Once in control, attackers can manipulate the device's functions, spy on users, or launch further attacks on the network. For example, researchers have demonstrated how hackers can take control of smart home systems, such as Amazon Ring cameras or Google Nest thermostats, to invade users' privacy or create disruptions.[4]

The implications of device hijacking extend beyond personal privacy concerns. In industrial environments, the hijacking of IoT devices can lead to operational disruptions or even physical damage. For instance, a hacker gaining control of an industrial IoT system used to manage critical infrastructure, such as power grids, water treatment plants, or manufacturing lines, could cause significant damage and economic losses. According to Lee et al., "the hijacking of IoT devices in critical infrastructure presents a national security risk, as it could lead to widespread disruptions in essential services." [5]

Data Privacy Concerns in IoT

IoT devices collect vast amounts of data, ranging from personal information to operational details in industrial systems. This data is often stored in the cloud or transmitted across networks, making it a prime target for cybercriminals. Many IoT devices lack robust encryption mechanisms, leaving sensitive data exposed during transmission. In addition, IoT devices are frequently connected to third-party services, further increasing the risk of data breaches.

A study conducted by the Ponemon Institute revealed that 58% of organizations using IoT devices experienced a security breach in 2020.[6] Data collected by IoT devices can include highly sensitive information, such as healthcare data from wearable devices or financial data from smart payment systems. The lack of proper encryption and security protocols makes these devices vulnerable to interception and misuse.

Mobile Devices: Increasing Targets for Cybercriminals

The widespread use of mobile devices for both personal and professional purposes has created a fertile ground for cyber threats. Mobile devices are now used for everything from online banking and shopping to managing sensitive corporate data. As a result, they have become attractive targets for cybercriminals looking to exploit vulnerabilities in mobile applications and operating systems.

One of the most prevalent threats to mobile devices is malware, which can be delivered through malicious apps, email attachments, or compromised websites. Mobile malware can steal personal information, track user activity, and even take control of the device's functions. In 2021, Check Point Research reported that 97% of organizations had encountered mobile malware in some form.[7] Mobile ransomware, which locks users out of their devices until a ransom is paid, has also become more common in recent years.

Another significant mobile threat is phishing, which targets users through SMS messages, email, or fraudulent apps. Mobile phishing attacks often involve tricking users into clicking on malicious links or downloading fake apps that steal their personal information. According to a report by Lookout, mobile phishing attacks increased by 37% in 2020 as cybercriminals adapted their strategies to target mobile users during the COVID-19 pandemic.[8]

Mobile Device Management (MDM) and Endpoint Security

To mitigate the risks associated with mobile threats, organizations are increasingly adopting Mobile Device Management (MDM)and Endpoint Security solutions. MDM allows IT administrators to remotely manage, monitor, and secure mobile devices used within the organization. This includes enforcing security policies, such as password protection, data encryption, and app management, to ensure that mobile devices comply with the organization's security standards.

Endpoint security extends beyond traditional antivirus software to include features like device tracking, remote wiping, and real-time threat detection. These tools are essential for protecting corporate data on mobile devices, especially in Bring Your Own Device (BYOD) environments where employees use personal devices to access company systems. According to Sinha, "mobile device management and endpoint security are critical in protecting corporate data from the growing range of threats targeting mobile platforms." [9]

Emerging Threats: 5G and IoT Integration

The rollout of 5G networks promises to revolutionize IoT by providing faster, more reliable connectivity and supporting billions of connected devices. However, the increased bandwidth and reduced latency of 5G networks also introduce new security challenges. As IoT devices become more interconnected, attackers may find new ways to exploit vulnerabilities in 5G infrastructure to launch attacks on IoT ecosystems.

In addition, the integration of 5G and IoT is expected to drive the development of smart cities, where IoT devices manage critical functions such as traffic control, public safety, and utilities. While these advancements offer significant benefits in terms of efficiency and convenience, they also create new attack vectors for cybercriminals. As highlighted by Meidan et al., "the integration of 5G and IoT in smart city infrastructures poses unique security risks, as attackers could disrupt essential services or gain access to sensitive data." [10]

Securing IoT and Mobile Environments

To address the emerging threats in the IoT and mobile landscapes, organizations should adopt a multi-layered security approach that includes device authentication, encryption, and regular software updates. For IoT devices, implementing strong device identity management systems can ensure that only authorized devices are allowed to connect to the network. Additionally, network segmentation can help limit the spread of malware or unauthorized access in the event that one device is compromised.

For mobile devices, encryption and two-factor authentication (2FA) are essential for protecting sensitive data. Regularly updating mobile operating systems and apps is also critical, as these updates often contain security patches that address newly discovered vulnerabilities. As Jain et al. note, "maintaining up-to-date software and enforcing strong authentication mechanisms are key strategies for securing mobile devices against the growing array of cyber threats."[11]

Conclusion

IoT devices offer a wealth of convenience and innovation, but they also create new attack surfaces that should be secured. Similarly, mobile devices, which now serve as critical tools for both personal and professional activities, are increasingly targeted by cybercriminals. To mitigate the risks posed by these emerging threats, organizations should implement robust security practices, including strong authentication, encryption, and regular updates, to protect their networks and data in the IoT and mobile landscapes.

Test Your Knowledge On The Cyber Threat Landscape

Q1: How have malware threats evolved over time?

Q2: What distinguishes ransomware from other forms of malware?

Q3: What is the economic impact of ransomware attacks?

Q4: How do social engineering attacks manipulate individuals?

Q5: What are Advanced Persistent Threats (APTs), and why are they dangerous?

Q6: How do nation-state actors use APTs in cyber warfare?

Q7: What is an insider threat, and how can it be mitigated?

Q8: What security challenges are posed by the increasing use of IoT devices?

Q9: What is the difference between phishing and spear-phishing?

Q10: How do attackers use DDoS attacks to disrupt services?

Q11: What are the potential consequences of data breaches?

Q12: How can organizations protect themselves against ransomware attacks?

Q13: How do cybercriminals use botnets in their attacks?

Q14: What is a zero-day exploit?

Q15: How does cybercrime on the dark web facilitate illegal activities?

ANSWERS

Q1: A: Malware has evolved from simple viruses and worms to sophisticated ransomware and polymorphic malware that can change its code to avoid detection.

Q2: A: Ransomware encrypts a victim's data and demands a ransom payment in exchange for restoring access, making it a financially motivated attack.

Q3: A: Ransomware attacks can result in significant financial losses due to downtime, ransom payments, legal fees, and reputational damage.

Q4: A: Social engineering attacks exploit human psychology to deceive victims into divulging sensitive information, often through phishing emails or fraudulent phone calls.

Q5: A: APTs are prolonged, targeted cyberattacks conducted by skilled adversaries who infiltrate a network and remain undetected for long periods to steal sensitive information.

Q6: A: Nation-state actors use APTs to conduct espionage, sabotage critical infrastructure, or gain a strategic advantage by infiltrating government or industry networks.

Q7: A: Insider threats arise from employees or contractors who misuse their access to harm the organization. Mitigation strategies include monitoring access and implementing privileged access management.

Q8: A: IoT devices often have limited security capabilities, making them easy targets for attackers who can exploit vulnerabilities to gain access to larger networks.

Q9: A: Phishing is a broad attack targeting many individuals, while spear-phishing is highly targeted and focuses on specific individuals, often using personalized information.

Q10: A: DDoS (Distributed Denial of Service) attacks overwhelm a system with traffic, causing it to become slow or unavailable to legitimate users.

Q11: A: Data breaches can lead to the theft of sensitive information, legal penalties, financial losses, and reputational damage for affected organizations.

Q12: A: Organizations can protect themselves by backing up data, educating employees about phishing, and using advanced endpoint protection to detect and block ransomware.

Q13: A: Cybercriminals use botnets—a network of infected devices—to launch large-scale attacks, including DDoS attacks and spam campaigns.

Q14: A: A zero-day exploit targets a previously unknown vulnerability, leaving organizations vulnerable because there is no available patch to protect against the attack.

Q15: A: The dark web provides a marketplace for cybercriminals to buy and sell stolen data, malware, and hacking services anonymously.

CHAPTER 4

CYBERSECURITY FOR INDIVIDUALS

Note: All references in this chapter are on page 398 - 399

We are going to focus on the individual cybersecurity safety. These are techniques for securing individual users and their data in an interconnected society. As we rely on digital platforms to do our everyday activities, we are the risk of personal data breaches, identity theft, and online fraud. We'll talk about ways to protect personal information, focused on everyday security habits like using strong, unique passwords, turning on multi-factor authentication (MFA), and making sure that devices and software are always up to date. We'll also stress how important it is to have good digital health and behave safely online by not clicking on sketchy links and only using networks you know you can trust. Also covered extensively will be the topic of password management and the ways in which such tools assist users in avoiding the usage of outdated or insecure credentials. Since home networks and individual devices are frequent entry points for hackers, this section also addresses how to protect them. Crimes such as social engineering, fraud, and theft are common. Here, you'll find information regarding these dangers and ways to be safe. After reading this chapter, you will have the knowledge and habits necessary to safeguard your online persona and conduct secure online transactions, making it more difficult for cybercriminals to target you.

4.1 Personal Data Protection Strategies

In the digital age, protecting personal data has become a critical aspect of cybersecurity for individuals. With the increasing reliance on online services and connected devices, personal data is more vulnerable to breaches, identity theft, and unauthorized access. Effective personal data protection strategies are essential for minimizing these risks, ensuring that individuals maintain control over their sensitive information while navigating the digital landscape.

The Importance of Personal Data Protection

Personal data includes a wide array of sensitive information, such as social security numbers, financial details, health records, and contact information. Cybercriminals often target this data for financial gain or use it to commit identity theft. According to the Federal Trade Commission (FTC), identity theft cases in the U.S. reached an all-time high in 2020, partly driven by the exploitation of personal data obtained through breaches and phishing attacks.[1]

The value of personal data extends beyond financial information. Various online services collect personal preferences, browsing habits, and even location data, making individuals more vulnerable to data misuse. As Gellman points out, "Personal data is the currency of the modern internet economy, but it also poses significant risks to privacy and security if not adequately protected."[2]

Strong Passwords and Multi-Factor Authentication

One of the most fundamental strategies for protecting personal data is the use of strong passwords. Weak or reused passwords are among the most common vulnerabilities exploited by cybercriminals. To create strong passwords, individuals should use a combination of uppercase and lowercase letters, numbers, and special characters. Avoid guessable password such as names or birthdates. Additionally, do not use one password for several online accounts. Use a unique password for each online account to ensure that other accounts remain secure if one password is compromised.

Individuals should enable multi-factor authentication (MFA) wherever possible to further enhance protection. MFA adds an additional layer of security by requiring users to provide a second form of verification—such as a one-time password (OTP), biometric scan, or physical token—before accessing an account. According to Verizon's Data Breach Investigations Report, "accounts protected by multi-factor authentication are significantly less likely to be compromised in a breach, as attackers should bypass multiple security barriers." [3]

A real-world example of the importance of MFA is the Twitter breach of 2020, in which attackers gained access to high-profile accounts through social engineering attacks. Many compromised accounts lacked MFA, making it easier for attackers to take control.[4] Enabling MFA on critical accounts, such as email and banking services, provides an additional safeguard against unauthorized access.

Data Encryption: Protecting Information in Transit and at Rest

Encryption is another vital strategy for safeguarding personal data. Encryption involves converting plaintext data into unreadable code that can only be decrypted by authorized parties with the correct key. This ensures that even if data is intercepted during transmission or stolen from a server, it remains protected from unauthorized access.

There are two primary types of encryption used to protect personal data: encryption in transit and encryption at rest. Encryption in transit protects data as it moves between devices, such as during online banking transactions or sending emails. Encryption at rest secures data stored on devices or servers, ensuring that sensitive information remains protected even if the device or server is compromised.

Major online services, such as Google, Apple and Meta, use end-to-end encryption to secure communications in applications like iMessage and WhatsApp, ensuring that only the sender and recipient can read the messages.[5] Encryption is also critical for protecting sensitive financial and healthcare data, as mandated by regulations like the General Data Protection Regulation (GDPR) and the Health Insurance Portability and Accountability Act (HIPAA).

Regular Software Updates and Patching

Cybercriminals often exploit software vulnerabilities to gain access to personal data. These vulnerabilities may exist in operating systems, applications, or devices, and attackers use malware or phishing attacks to exploit them. Regularly updating software and applying security patches is essential for closing these security gaps and protecting personal information.

According to a report by Ponemon Institute, "nearly 60% of data breaches could be traced back to unpatched software vulnerabilities."[6] By keeping devices and applications up to date, individuals can significantly reduce their exposure to cyber threats. Many software providers offer automatic updates. This ensures that users receive the latest security patches without needing to take manual action.

Secure Backups: Protecting Against Ransomware and Data Loss

Regular data backups are a crucial component of personal data protection strategies. In the event of a cyberattack, particularly a ransomware attack, having a secure backup ensures that individuals can recover their data without paying a ransom. Backups are also important in cases of hardware failure, accidental deletion, or natural disasters that may lead to data loss. Backing up individual phones and personal computers is essential.

Experts recommend using both local and cloud-based backups to create redundancy. For example, a local backup can be stored on an external hard drive, while a cloud backup ensures that data is stored offsite and protected from physical damage to devices. According to a report by Kaspersky Lab, "Regular, secure backups are one of the most effective ways to recover from ransomware attacks without significant data loss."[7]

Privacy Settings and Data Minimization

Many online services collect vast amounts of personal data, often more than is necessary for their operation. Individuals can protect their privacy by adjusting the privacy settings on websites, apps, and devices to limit the amount of data collected. For example, users can disable location tracking, reduce data sharing with third-party apps, and limit the collection of personal preferences.

Data minimization is another important strategy, where individuals only provide the minimum amount of personal information required to use a service. This can reduce the risk of sensitive data being exposed in the event of a breach. As Schneier argues, "The less data an individual shares online, the less valuable they become as a target for cybercriminal."[8]

Phishing Awareness and Safe Browsing Practices

Phishing attacks remain one of the most common methods cybercriminals use to steal personal data. These attacks typically involve emails or messages that appear to come from legitimate sources, tricking users into clicking malicious links or downloading infected attachments. Once the victim interacts with the phishing email, attackers may steal login credentials or install malware on the device.

To protect against phishing, individuals should exercise caution when opening unsolicited emails or clicking on unfamiliar links. Anti-phishing software and browser extensions can also help detect and block malicious websites before they cause harm. Additionally, secure browsing practices, such as browsing only websites that use HTTPS and avoiding public Wi-Fi for sensitive transactions, further reduce the data interception risk.

A notable example of a large-scale phishing attack occurred in 2016 when attackers targeted Gmail users with fake emails posing as Google Drive notifications. These emails contained a link to a fraudulent website designed to steal login credentials. Millions of users were affected, underscoring the importance of verifying the legitimacy of email communications before taking action.[9]

Conclusion

In an increasingly connected world, personal data is more vulnerable to theft and misuse than ever before. By adopting robust personal data protection strategies such as using strong passwords, enabling multi-factor authentication, encrypting sensitive data, and staying vigilant against phishing attacks, individuals can safeguard their information and reduce the risk of becoming victims of cybercrime. Personal data protection remains a critical aspect of cybersecurity for individuals, requiring ongoing attention and proactive measures.

4.2 Secure Online Behavior and Digital Hygiene

In an era where cyberattacks are becoming more prevalent and sophisticated, maintaining secure online behavior and practicing good digital hygiene are essential for protecting personal information and avoiding security breaches. Cybercriminals often exploit user behavior, targeting weak passwords, unpatched systems, and careless online habits. By adopting secure online practices and maintaining proper digital hygiene, individuals can significantly reduce their vulnerability to cyber threats and safeguard their digital presence.

The Concept of Digital Hygiene

Digital hygiene refers to the set of practices that help individuals maintain the security, privacy, and functionality of their digital environments. It encompasses a wide range of activities, from updating software regularly to managing passwords effectively. As emphasized by Gordon and Loeb, "digital hygiene is a proactive approach to cybersecurity that focuses on preventing problems before they arise, much like personal hygiene prevents illness."[1]

Practicing good digital hygiene is critical because even minor lapses can open the door to cyberattacks. For example, failing to update software regularly or using weak passwords increases the risk of malware infections or unauthorized access to personal accounts. Digital hygiene is not only about protecting devices but also about safeguarding personal information and online behavior.

Managing Privacy Settings and Data Sharing

Individuals often share personal information online without fully understanding how it can be used or misused by third parties. Social media platforms, apps, and websites often collect more data than necessary for their operation and this creates privacy risks. Adjusting privacy settings to limit data collection and sharing is key to secure online behavior.

For instance, individuals can disable location tracking on mobile apps, limit access to contacts and photos, and prevent third-party apps from accessing personal information. As Schneier notes, "the more data individuals share online, the more vulnerable they become to data breaches and identity theft."[2]

Avoiding Public Wi-Fi for Sensitive Transactions

Public Wi-Fi networks are convenient but often insecure, making them a prime target for cybercriminals looking to intercept data. When using public Wi-Fi, sensitive transactions such as online banking, shopping, or accessing private accounts, should be avoided. Virtual Private Networks (VPNs) offer a secure solution for individuals who need to use public Wi-Fi while protecting their online activities. A VPN encrypts the user's internet traffic, ensuring that even if the data is intercepted, it cannot be read by attackers.

According to a study by Norton, "nearly 60% of public Wi-Fi users fail to take basic security precautions, such as using a VPN or avoiding sensitive transactions, leaving them vulnerable to attacks."[3]. By using a VPN, individuals can protect their data from eavesdropping and man-in-the-middle attacks while using public networks.

Cybersecurity Awareness and Continuous Learning

Maintaining secure online behavior requires individuals to stay informed about the latest cybersecurity threats and trends. Cybercriminals constantly evolve their tactics, making it essential for individuals to educate themselves about new risks and best practices. Cybersecurity awareness training programs are commonly offered by employers, but individuals can also take advantage of free online resources and workshops to improve their knowledge.

As cyber threats become more sophisticated, continuous learning is key to adapting to new challenges. As Anderson and Moore argue, "Cybersecurity is an ongoing process that requires vigilance, adaptation, and education, particularly as the digital landscape continues to evolve."[4]

Conclusion

Secure online behavior and good digital hygiene are essential for protecting personal data and reducing the risk of cyberattacks. By managing passwords effectively, enabling multi-factor authentication, updating software regularly, and practicing safe browsing habits, individuals can create a secure digital environment. Maintaining strong digital hygiene practices will be crucial in safeguarding personal information and preventing security breaches.

4.3 Password Management and Authentication Best Practices

In today's digital landscape, password management and authentication best practices are essential to maintaining the security of personal data and preventing unauthorized access to online accounts. Passwords are the primary method of securing accounts, but poor password habits, such as reusing weak passwords, create vulnerabilities that cybercriminals can exploit. The implementation of robust authentication methods, such as multi-factor authentication (MFA), adds additional layers of security, further protecting against cyberattacks. By following best practices in password management and authentication, individuals can significantly reduce the risk of data breaches and identity theft.

The Role of Strong Passwords in Security

Passwords serve as the first line of defense against unauthorized access to personal accounts and sensitive data. A strong password is one that is unique, complex, and difficult to guess. It should include a combination of uppercase and lowercase letters, numbers, and symbols. Passwords that rely on simple patterns or personal information, such as birthdates or names, are more easily cracked by attackers using brute-force techniques or dictionary attacks. As Florencio and Herley explain, "Users often underestimate the importance of password strength, opting for convenience over security."[1]

A study by SplashData found that the most common passwords used in 2020 were still weak, with "123456" and "password" topping the list.[2] These predictable choices make it easy for attackers to gain access to accounts, particularly when passwords are reused across multiple platforms. To combat this, experts recommend using password managers to generate and store strong, random passwords for each account. Password managers eliminate the need to memorize multiple passwords and reduce the likelihood of using weak or reused credentials. It worth noting that Password managers can be compromised too.

Password Reuse: A Common Vulnerability

One of the most significant risks in password management is password reuse. When individuals use the same password across multiple accounts, a breach of one service can expose all other accounts to unauthorized access. For example, if a user's email account is compromised due to password reuse, attackers can use the same password to gain access to other services, such as banking or social media accounts.

According to a report by Verizon, "81% of data breaches are caused by weak or reused passwords."[3] Password reuse is particularly dangerous because cybercriminals often use credential stuffing techniques, where they attempt to use stolen credentials from one breach to access other accounts. To prevent this, individuals should use unique passwords for each account and regularly change them, particularly for high-value accounts such as those associated with banking or healthcare services.

Multi-Factor Authentication: Strengthening Password Security

While strong passwords are critical, they are not foolproof. Multi-factor authentication (MFA) provides an additional layer of security by requiring users to provide more than just a password to access their accounts. MFA typically combines something the user knows (a password) with something they have (a mobile device) or something they are (biometrics). This dual verification process significantly reduces the likelihood of unauthorized access, even if the password is compromised.

MFA methods include one-time passwords (OTPs) sent via SMS or email, authenticator apps, and biometric authentication methods such as fingerprint or facial recognition. As suggested by Anderson and Moore, "MFA mitigates the risk of password compromise by ensuring that attackers should bypass multiple security barriers to gain access." [4]

The effectiveness of MFA was demonstrated in the Google Cloud Platform breach in 2019. Attackers compromised user credentials but were unable to access the accounts due to the activation of MFA, which required a second form of authentication beyond the password.[5] For sensitive accounts, such as those associated with financial services, healthcare records, and personal communications, enabling MFA is essential for protecting against unauthorized access.

Password Policies and Best Practices

Organizations often implement password policies to ensure that users maintain strong security habits. These policies typically dictate the complexity, length, and expiration of passwords. For instance, many applications require passwords to be at least 12 characters long and to include a mix of letters, numbers, and symbols. Additionally, regular password expiration policies require users to change their passwords every 60 or 90 days, reducing the risk of long-term exposure in the event of a breach.

However, password policies should balance security with usability. Overly complex or frequent password changes can lead to poor user practices, such as writing passwords down or reusing old passwords. According to Florencio and Herley, "forcing users to change their passwords too frequently can backfire, as users are likely to adopt weaker passwords or reuse old ones to avoid the inconvenience." [6] Therefore, organizations should carefully design their password policies to promote security and user compliance and convenience.

The Role of Biometrics in Authentication

Biometric authentication, which uses physical or behavioral characteristics such as fingerprints, facial recognition, or voice recognition, offers a secure and convenient alternative to traditional password-based authentication. Biometrics are difficult to replicate, providing a high level of security for personal accounts. Many modern devices, such as smartphones and laptops, now include biometric features, making it easier for individuals to use these technologies for daily authentication.

For instance, Apple's Face ID and Touch ID use facial recognition and fingerprint scanning to authenticate users, allowing for secure access to devices and applications without the need for a password. Similarly, Microsoft Windows Hello offers biometric authentication for Windows devices, enhancing security while improving user convenience.[7] Biometrics are increasingly used in financial services as well, where banks implement fingerprint or facial recognition to verify transactions and protect customer accounts.

According to Jain et al., "biometric authentication is an essential component of future-proof security, offering enhanced protection against credential theft and unauthorized access."[8] However, biometrics should be used in combination with other security measures, such as MFA, to provide comprehensive protection.

The Future of Authentication: Beyond Passwords

As cyber threats continue to evolve, the limitations of passwords are becoming more apparent. Passwords are prone to human error, such as weak choices or reuse, and are vulnerable to modern attack techniques like phishing and brute-force attacks. The future of authentication is likely to move beyond passwords, with passwordless authentication becoming more prevalent.

Passwordless authentication methods, such as biometric verification or public key cryptography, allow users to authenticate themselves without the need for a password. For example, FIDO2 (Fast Identity Online) standards enable passwordless logins using biometric data or cryptographic keys stored on a secure device. This eliminates the risk of password theft while maintaining a high level of security.

As Stajano and Anderson explain, "Passwordless authentication represents a paradigm shift in cybersecurity, reducing the reliance on human-generated passwords and improving overall security outcomes."[9] By integrating passwordless technologies with traditional authentication methods, organizations can provide a more secure and user-friendly experience.

Conclusion

Password management and authentication are critical components of personal cybersecurity. By following best practices such as using strong, unique passwords, enabling multi-factor authentication, and implementing password managers, individuals can significantly enhance their security and reduce the risk of unauthorized access.

4.4 Protecting Personal Devices and Home Networks

As more individuals integrate personal devices and smart home technologies into their daily lives, protecting personal devices and home networks has become an essential component of cybersecurity. Devices such as smartphones, laptops, tablets, and IoT (Internet of Things) gadgets, coupled with home networks, are increasingly targeted by cybercriminals aiming to steal personal data or disrupt operations. Implementing strong security measures for both devices and home networks is critical to safeguarding personal information and maintaining privacy in the digital age.

The Risks of Unsecured Devices and Networks

Unsecured personal devices and home networks are prime targets for cybercriminals. Personal devices often contain sensitive information such as banking details, personal photos, and documents. Additionally, home networks are frequently used to access online services, manage smart home systems, and store data on cloud platforms. If left unprotected, these networks can be exploited by attackers to steal information, monitor activities, or launch attacks on other systems.

According to Symantec's Internet Security Threat Report, "home networks are increasingly being targeted by cybercriminals as more individuals rely on connected devices without implementing basic security measures."[1] The risks of unsecured devices and networks extend beyond personal inconvenience; they can also have serious financial and privacy implications, particularly if personal data is compromised.

Securing Personal Devices

Securing personal devices such as smartphones, laptops, and tablets begins with the basics: enabling password protection, automatic updates, and encryption. Device passwords should be complex, using a combination of letters, numbers, and symbols, and devices should be locked when not in use. Additionally, enabling multi-factor authentication (MFA)provides an extra layer of protection, requiring a second form of verification before accessing sensitive accounts.

Regular software updates are critical for securing personal devices. Software vulnerabilities are often exploited by malware or attackers, and patches provided by manufacturers address these vulnerabilities. According to a report by the Ponemon Institute, "unpatched software vulnerabilities remain one of the leading causes of cyberattacks on personal devices."[2] By keeping operating systems and applications up to date, individuals can reduce their exposure to these risks.

Encryption is another essential security feature for protecting personal devices. Encryption converts data into unreadable code that can only be decrypted by authorized users. Both Apple and Android devices offer encryption options to protect stored data, ensuring that even if the device is lost or stolen, the data remains inaccessible to unauthorized users.[3] Using encryption on laptops and other storage devices, such as external hard drives, provides further protection for sensitive data.

Protecting Home Networks

Home networks, particularly those using Wi-Fi, are frequently targeted by cybercriminals due to their widespread use and, often, poor security practices. Protecting a home network begins with securing the Wi-Fi router, which acts as the central hub for all connected devices. Routers should be protected with strong, unique passwords, and the default login credentials, often set by the manufacturer, should be changed immediately after installation. According to Kaspersky Lab, "default router passwords are easily discovered by cybercriminals and are one of the most common entry points for network intrusions."[4]

Enabling network encryption is also critical for securing home Wi-Fi. The most secure form of encryption available for home networks is WPA3 (Wi-Fi Protected Access 3), which provides stronger protection than its predecessor WPA2. WPA3 encryption ensures that data transmitted over the network is protected from eavesdropping and tampering. Additionally, disabling remote access to the router and enabling firewall settings further strengthen network security by preventing unauthorized access from external sources.

Network segmentation is another useful strategy for protecting home networks. By creating separate networks for different devices, such as one network for personal devices and another for smart home gadgets or guest users, individuals can reduce the risk of a single compromised device that would jeopardize the entire network. For example, IoT devices like smart thermostats or security cameras can be isolated on their own network, minimizing the potential impact of an attack on more critical systems.

IoT Device Security

The rapid growth of the Internet of Things (IoT) has introduced additional security challenges for home networks. IoT devices, such as smart speakers, smart refrigerators, and security cameras, are often less secure than traditional computing devices, making them attractive targets for cybercriminals. Many IoT devices are shipped with default passwords and minimal security features, which users often fail to change or update.

According to Gartner, "By 2025, it is estimated that IoT devices will account for over 75 billion connected devices, creating a significant challenge for securing home networks."[5] To protect IoT devices, users should change default passwords upon installation, regularly check for and apply firmware updates, and disable features that are not necessary, such as remote access or data sharing. Additionally, placing IoT devices on a segmented network can further isolate them from more sensitive devices, such as laptops or smartphones.

Virtual Private Networks (VPNs) for Secure Connectivity

Using a Virtual Private Network (VPN) is one of the most effective ways to protect personal devices and home networks from unauthorized access, especially when using public Wi-Fi networks. A VPN encrypts all internet traffic, ensuring that any data transmitted between devices and the internet is unreadable to attackers. VPNs are particularly useful when individuals are working remotely or accessing sensitive information through unsecured networks.

As highlighted by NordVPN, "using a VPN significantly reduces the risk of man-in-the-middle attacks, where cybercriminals intercept and manipulate data traveling between users and websites."[6] By using a VPN, individuals can ensure that their online activities remain private and protected from prying eyes, whether at home or on the go.

Parental Controls and Monitoring for Family Safety

For families with children, implementing parental controls and monitoring tools can help protect younger users from inappropriate content, online predators, and malicious websites. Most modern routers offer built-in parental controls, allowing parents to restrict access to certain websites or limit the time spent online. Additionally, monitoring software can provide insights into children's online behavior and alert parents to any potential security or privacy concerns.

According to a report by Common Sense Media, "Parents should take an active role in monitoring their children's online activities and educating them about safe internet use."[7] Teaching children about the importance of online privacy, avoiding suspicious links, and practicing good password habits can help foster secure online behavior from a young age.

The Future of Home Network Security

The future of home network security will likely involve the integration of artificial intelligence (AI) and machine learning (ML) technologies. These tools can help identify patterns of abnormal behavior across devices, detect potential intrusions in real time, and automatically block suspicious activity before it compromises the network. AI-driven security solutions are expected to play an increasingly important role in defending personal devices and home networks against sophisticated cyberattacks.

As Sinha and Rao argue, "AI and ML will be critical in developing adaptive security measures that respond dynamically to emerging threats, particularly in the context of the IoT and smart home environments."[8] The adoption of advanced security technologies will be essential for keeping pace with the growing complexity of cyber threats in the digital age.

Conclusion

Protecting personal devices and home networks is essential for maintaining privacy, security, and the integrity of sensitive information in an increasingly connected world. By implementing strong security practices such as enabling password protection, using encryption, securing routers, and regularly updating software, individuals can reduce their risk of falling victim to cyberattacks. As IoT devices and smart home technologies continue to proliferate, adopting proactive security measures will be crucial to safeguarding personal data and maintaining secure home networks.

4.6 Recognizing and Avoiding Common Cyber Scams

Cyber scams have become increasingly prevalent in the digital world, targeting individuals through various deceptive techniques. These scams range from phishing emails to online fraud, with cybercriminals continuously evolving their methods to exploit users' vulnerabilities. Understanding how to recognize and avoid these scams is essential for individuals to protect their personal data, financial information, and overall digital security. By adopting critical thinking, skepticism, and best practices in cybersecurity, users can minimize the risks posed by cyber scams.

Phishing Scams: A Widespread Threat

One of the most common cyber scams is phishing, where attackers impersonate legitimate entities, such as banks, government agencies, or online services, to trick users into providing sensitive information.

These scams typically arrive in the form of emails or text messages, often with urgent language designed to provoke immediate action. According to a report by the Anti-Phishing Working Group (APWG), "phishing attacks have steadily increased over the past decade, with a significant rise during global events such as the COVID-19 pandemic."[1]

Phishing messages usually contain links to fraudulent websites that appear identical to legitimate ones. Victims are prompted to enter personal information, such as login credentials or credit card numbers, which are then harvested by the attackers. To avoid falling victim to phishing, individuals should scrutinize the sender's email address, hover over links to verify their authenticity, and avoid clicking on suspicious attachments. Many email providers and web browsers now include anti-phishing tools that warn users when they encounter potentially malicious content.

A well-known example of a phishing scam occurred in 2016 when attackers targeted users of Gmail with fake notifications from Google Drive. The scam involved emails that appeared to come from Google, asking users to click a link to view shared documents. Instead, the link led to a fake login page designed to steal users' credentials. This incident underscores the importance of verifying the authenticity of email communications before taking action.[2]

Tech Support Scams: Exploiting Fear and Confusion

Another common scam involves tech support fraud, where scammers pose as technical support agents from well-known companies, such as Microsoft or Apple. Victims are told that their computer has been infected with malware or is experiencing a critical error, and they are urged to grant remote access to fix the issue. In reality, the scammers use this access to steal personal information, install malware, or extort money from the victim.

Tech support scams often target older individuals or those less familiar with technology, exploiting their fear and confusion. According to a report by the Federal Trade Commission (FTC), "tech support scams accounted for over $147 million in losses in 2020, with many victims unaware that legitimate companies would never initiate unsolicited technical support calls."[3] To avoid falling victim to these scams, individuals should remember that reputable companies do not proactively offer tech support via unsolicited phone calls or pop-up messages.

A notable case of tech support fraud occurred in 2018 when scammers posing as Microsoft support called individuals and convinced them to grant remote access to their computers. The scammers then demanded payment to "resolve" the fabricated issues, resulting in millions of dollars in losses. The incident highlights the importance of skepticism when dealing with unsolicited tech support offers.[4]

Online Shopping Scams: The Rise of Fake E-commerce Sites

The growth of online shopping has led to an increase in e-commerce scams, where fraudulent websites offer products at attractive prices but fail to deliver them. These fake e-commerce sites often mimic legitimate online retailers, complete with stolen product images and customer reviews. Victims are lured in by the promise of deep discounts, only to discover that their personal and financial information has been stolen.

To avoid falling victim to online shopping scams, individuals should verify the legitimacy of websites before making a purchase. This includes checking for secure payment methods (such as HTTPS in the URL), reading customer reviews, and researching the company's background. According to Norton, "shopping on well-known, reputable websites and avoiding deals that seem too good to be true are key strategies for avoiding online shopping scams."[5]

In 2020, a surge in online shopping during the COVID-19 pandemic led to a wave of fake e-commerce sites offering everything from personal protective equipment (PPE) to consumer electronics. Many victims paid for products that were never delivered or received counterfeit goods. This example underscores the need for consumers to remain vigilant when shopping online, particularly during times of increased demand.[6]

Romance Scams: Exploiting Emotional Vulnerabilities

Romance scams are a particularly insidious form of cyber fraud, where scammers build emotional connections with victims through online dating platforms or social media. Over time, the scammer gains the victim's trust, then invents a financial emergency or travel dilemma, and ask the victim to send money. These scams often continue for months, with the scammer making repeated requests for financial assistance.

Romance scams are emotionally and financially devastating, as victims often develop genuine feelings for the scammer. According to the FBI's Internet Crime Complaint Center (IC3), "romance scams caused over $304 million in losses in 2020, making them one of the most financially damaging types of cyber scams."[7] Victims are often too embarrassed to report the scam, allowing scammers to continue their schemes with impunity.

To avoid romance scams, individuals should be cautious when forming online relationships with people they have never met in person. Red flags include requests for money, reluctance to meet in person or via video chat, and inconsistent or evasive answers to personal questions. Scammers often use fake photos and fabricated stories. It's essential to verify the identity of online acquaintances before becoming emotionally or financially invested.

Investment and Cryptocurrency Scams

Investment scams, including cryptocurrency fraud, have become more common as interest in digital currencies and online trading platforms grows. These scams typically involve promises of guaranteed returns or insider knowledge, enticing victims to invest large sums of money. Once the victim transfers funds, the scammer disappears, leaving the victim with significant financial losses.

Cryptocurrency scams are particularly appealing to scammers because of the anonymous nature of Blockchain transactions, making it difficult for victims to trace their funds. According to Chainalysis, "Cryptocurrency scams accounted for nearly $14 billion in stolen funds in 2021, with scammers taking advantage of the rising popularity of digital currencies." [8] Common tactics include fake investment platforms, Ponzi schemes, and phishing attacks targeting cryptocurrency wallets.

To protect against investment scams, individuals should be skeptical of promises of high returns with little risk. Conducting thorough research on investment opportunities and using regulated financial platforms can help reduce the risk of falling victim to these scams. Additionally, individuals should verify the legitimacy of cryptocurrency exchanges and avoid sharing private keys or wallet information with untrusted sources.

Avoiding Social Engineering Scams

Social engineering scams manipulate individuals into divulging personal information or performing actions that compromise security. These scams rely on psychological manipulation, often exploiting emotions such as fear, trust, or urgency. Common social engineering tactics include impersonation, baiting, and pretexting—where the scammer creates a false scenario to justify their request for information.

To defend against social engineering scams, individuals should be cautious about sharing sensitive information with anyone, especially if the request comes through unsolicited phone calls or emails. Verifying the identity of the requester and questioning the legitimacy of the scenario can help prevent falling victim to these scams. As Mitnick and Simon argue, "the human element remains the weakest link in cybersecurity, with social engineering exploiting trust and emotions to bypass technical safeguards."[9]

Conclusion

Cyber scams are diverse and continually evolving. By understanding the most common types of scams, such as phishing, tech support fraud, online shopping scams, and romance scams, individuals can adopt best practices to protect themselves. Critical thinking, skepticism, and adherence to cybersecurity principles, such as using strong authentication and verifying the legitimacy of communications, are key to avoiding cyber scams.

Test Your Knowledge on Cybersecurity for Individuals

Q1: What are the most effective strategies for protecting personal data online?

Q2: How does secure online behavior contribute to cybersecurity?

Q3: What role do password managers play in personal cybersecurity?

Q4: What are some best practices for creating and managing secure passwords?

Q5: How does multi-factor authentication (MFA) enhance security?

Q6: What steps can individuals take to protect their personal devices from cyber threats?

Q7: How can individuals secure their home networks against cyberattacks?

Q8: What are the risks associated with public Wi-Fi, and how can they be mitigated?

Q9: How can individuals recognize and avoid common cyber scams?

Q10: Why is it important to keep personal devices and software up to date?

Q11: What are the most common types of personal cyber threats?

Q12: How does using encryption protect personal data?

Q13: What should individuals do if they suspect their personal information has been compromised?

Q14: How can individuals safeguard their social media accounts from cyber threats?

Q15: What are the risks of oversharing personal information online?

ANSWERS

Q1: A: Strategies include using strong, unique passwords, enabling multi-factor authentication, regularly updating software, and avoiding sharing sensitive information on unsecured platforms.

Q2: A: Secure online behavior, such as avoiding suspicious links and verifying the legitimacy of websites, helps prevent individuals from falling victim to phishing or malware attacks.

Q3: A: Password managers store complex, unique passwords for each account, reducing the risk of password reuse and making it easier to use strong passwords.

Q4: A: Best practices include using long, complex passwords with a mix of characters, avoiding common phrases, and changing passwords regularly.

Q5: A: MFA requires two or more verification methods, such as a password and a fingerprint, making it significantly harder for attackers to gain unauthorized access.

Q6: A: Individuals should install security software, enable automatic updates, use firewalls, and avoid downloading apps from untrusted sources.

Q7: A: Securing a home network involves changing default router passwords, using encryption (WPA3), disabling remote management, and regularly updating firmware.

Q8: A: Public Wi-Fi is often unsecured, making it easy for attackers to intercept data. Using a virtual private network (VPN) mitigates these risks by encrypting the user's internet connection.

Q9: A: Common cyber scams, such as phishing emails or fake tech support calls, can be avoided by verifying sources, being skeptical of unsolicited requests, and checking for suspicious URLs.

Q10: A: Updates often include security patches for newly discovered vulnerabilities, making it essential to keep devices and software current to protect against exploits.

Q11: A: Common threats include identity theft, phishing, malware infections, and social engineering attacks that target individuals' personal information.

Q12: A: Encryption secures data by converting it into unreadable code, ensuring that only authorized users with the decryption key can access the information.

Q13: A: They should immediately change passwords, monitor financial accounts for suspicious activity, and consider placing fraud alerts on credit reports.

Q14: A: Safeguarding social media involves using strong, unique passwords, enabling MFA, reviewing privacy settings, and being cautious of unsolicited messages or friend requests.

Q15: A: Oversharing can lead to identity theft, targeted scams, or social engineering attacks, as attackers can use publicly available information to exploit individuals.

CHAPTER 5

ORGANIZATIONAL CYBERSECURITY

Note: All references in this chapter are on page 399 - 400

This chapter covers how organizations and companies could create all-encompassing plans to guard against cyberattacks. We will start with a general review of the key components such as risk assessment, threat detection and incident response planning that support a strong cybersecurity program. You will learn about several risk management systems that enable companies to recognize and rank their most important assets as well as techniques for applying security policies that fit both legal criteria and corporate goals. One of the main priorities is employee training since human mistake usually contributes to security lapses. This chapter will walk you through how to put successful training initiatives into action so that employees may identify and lessen risks. We will also discuss the need of privileged access management (PAM) and how restricted access to important systems could stop insider threats. By the end of this chapter, you will know how companies could create robust defenses while guaranteeing company continuity and so reducing the financial consequences of a cybercrime. Essential for surviving and rebuilding from cyberattacks, incident response techniques and business continuity planning will also be taught to you.

5.1 Develop a Comprehensive Cybersecurity Strategy

In today's digital age, organizations face an ever-growing range of cyber threats, from ransomware attacks to data breaches and insider threats. A comprehensive cybersecurity strategy is essential for protecting sensitive data, maintaining business continuity, and ensuring compliance with industry regulations. Such a strategy requires a holistic approach to addressing technical, organizational, and human factors. A robust cybersecurity framework involves:

- Identifying vulnerabilities.
- Implementing security controls.
- Continuously monitoring for threats.
- Fostering a culture of security awareness within the organization.

Understanding the Importance of Cybersecurity Strategy

Cybersecurity is no longer a technical issue confined to the IT department. It is a critical aspect of organizational risk management that potentially affect reputation, financial stability, and legal compliance. According to Gartner, "the average cost of a data breach is now over $4 million, making cybersecurity a top priority for organizations of all sizes."[1] A comprehensive strategy is essential not only for preventing attacks but also for minimizing damage when breaches do occur.

The primary goal of a cybersecurity strategy is to protect the confidentiality, integrity, and availability of data and systems—often referred to as the CIA triad. Confidentiality ensures that sensitive information is accessible only to authorized users, integrity ensures that data is accurate and unaltered, and availability ensures that systems remain operational and accessible when needed. A failure to address any of these pillars can lead to severe consequences for an organization, including loss of customer trust, regulatory penalties, and financial loss.

Assessing Risks and Identifying Vulnerabilities

The first step in developing a comprehensive cybersecurity strategy is conducting a thorough risk assessment. This process involves identifying the organization's most valuable assets, such as customer data, intellectual property, or operational systems, and assessing the risks associated with those assets. The risk assessment should consider both internal and external threats, including malicious insiders, cyber criminals, and nation-state actors. According to NIST's Cybersecurity Framework, "a comprehensive risk assessment should address potential threats, vulnerabilities, and the likelihood of exploitation."[2]

Once risks have been identified, the organization should also identify specific vulnerabilities within its systems. Vulnerabilities can arise from outdated software, weak passwords, unpatched systems, or misconfigured network devices. Penetration testing, also known as ethical hacking, can help organizations identify these vulnerabilities by simulating cyberattacks and assessing the effectiveness of current defenses. Organizations can prioritize the remediation of high-risk areas by understanding where weaknesses exist.

Implement Security Controls

A critical component of any cybersecurity strategy is the implementation of security controls that protect the organization's assets and mitigate identified risks. Security controls fall into three broad categories: preventive, detective, and corrective.

- **Preventive controls** are designed to stop cyberattacks before they happen. These include firewalls, antivirus software, intrusion prevention systems (IPS), and access control mechanisms. Preventive measures aim to limit unauthorized access to systems and prevent the introduction of malware into the network. For example, multi-factor authentication (MFA) requires users to verify their identity using two or more factors, making it more difficult for attackers to gain access to accounts.

- **Detective controls** help organizations identify and respond to security incidents when they occur. Intrusion detection systems (IDS), security information and event management (SIEM) platforms, and log monitoring are all detective controls that provide real-time visibility into network activity. These tools alert security teams to potential threats, allowing for a rapid response to contain the attack.

- **Corrective controls** are implemented after a security incident to mitigate its impact and restore normal operations. Examples include disaster recovery plans, data backups, and incident response teams. These controls are essential for minimizing downtime and reducing the long-term effects of a cyberattack.

According to ISO/IEC 27001, a leading standard for information security management, "the effective implementation of security controls requires a layered approach that integrates multiple defensive measures to protect against a wide range of threats."[3] By implementing controls across different layers of the network, organizations can create a more resilient defense against cyberattacks.

Continuous Monitoring and Threat Detection

Cyber threats are constantly evolving, and new vulnerabilities are discovered regularly. As a result, a comprehensive cybersecurity strategy should include continuous monitoring and threat detection capabilities. Security Operations Centers (SOCs) play a crucial role in monitoring an organization's systems for signs of malicious activity. SOCs use tools such as SIEM platforms, which aggregate and analyze logs from various sources to detect anomalies or suspicious behavior.

Proactive threat detection involves identifying potential security incidents before they result in damage. Techniques such as behavioral analysis, machine learning, and threat intelligence can help organizations identify patterns of abnormal behavior and respond to threats more effectively. For example, a machine learning-based SIEM system can flag unusual login attempts from unfamiliar locations, allowing security teams to investigate further. As Saxena et al. note, "AI-powered threat detection tools are becoming increasingly important in detecting advanced persistent threats (APTs) and zero-day attacks that evade traditional defenses."[4]

Incident Response Plan

No cybersecurity strategy is complete without a well-defined incident response plan. An incident response plan outlines the steps that an organization will take in the event of a cyberattack, from identifying the breach to containing the damage and recovering operations. A successful response plan minimizes the impact of the attack, reduces downtime, and ensures that the organization can continue to operate.

SANS Institute recommends that incident response plans include the following phases: preparation, identification, containment, eradication, recovery, and lessons learned.[5] These phases ensure that organizations are prepared to respond quickly to incidents while also improving their security posture for the future. Post-incident analysis is particularly important, as it allows the organization to identify the root cause of the breach and implement additional security measures to prevent similar attacks in the future.

Foster a Culture of Security Awareness

While technical controls are essential for cybersecurity, human behavior plays a significant role in preventing cyberattacks. Social engineering attacks, such as phishing, exploit human vulnerabilities, making it critical for organizations to foster a culture of security awareness among employees. Regular security training programs can help employees recognize phishing emails, avoid clicking on suspicious links, and report potential threats to the IT department.

According to Verizon's 2021 Data Breach Investigations Report, "human error is a contributing factor in over 85% of data breaches."[6] This underscores the importance of employee training in cybersecurity. Organizations can reduce the risk of successful attacks by educating employees about common cyber threats and encouraging them to adopt secure online behaviors.

Ensure Compliance with Industry Regulations

In addition to protecting organizational assets, a comprehensive cybersecurity strategy should ensure compliance with industry-specific regulations and standards. Regulatory frameworks such as GDPR (General Data Protection Regulation), HIPAA (Health Insurance Portability and Accountability Act), and PCI DSS (Payment Card Industry Data Security Standard) impose strict requirements for the protection of personal data and the reporting of data breaches. Failure to comply with these regulations can result in significant fines and reputational damage.

As noted by Kesan and Hayes, "compliance with cybersecurity regulations is not only a legal obligation but also a competitive advantage, as customers increasingly prioritize the security of their data."[7] Regular compliance audits and security assessments help organizations verify that they meet regulatory requirements and maintain robust security practices.

Conclusion

Developing a comprehensive cybersecurity strategy requires a proactive, multi-faceted approach that integrates risk assessment, security controls, continuous monitoring, and employee training. Organizations can protect their assets, maintain business continuity, and comply with regulatory requirements by implementing robust defenses and preparing for potential incidents. Organizations should remain vigilant, regularly updating their cybersecurity strategies to address emerging risks and vulnerabilities.

5.2 Risk Management Frameworks

Effective risk assessment and management frameworks are critical to the development and implementation of a comprehensive cybersecurity strategy. Organizations are constantly exposed to various cyber threats. Risk assessment frameworks help organizations identify, evaluate, and prioritize risks, while risk management frameworks offer a structured approach to mitigating those risks. These frameworks provide organizations with the tools to protect their assets, comply with regulatory requirements, and ensure business continuity.

The Role of Risk Assessment in Cybersecurity

Risk assessment helps organizations understand the vulnerabilities that exist in their systems and the likelihood and impact of a cyberattack. According to NIST's Risk Management Framework, "risk assessment is a critical component of an effective cybersecurity program, as it allows organizations to allocate resources to the areas of highest risk."[1]

Risk assessment enables organizations to prioritize their security efforts and allocate resources to areas that pose the greatest risk. For example, an organization might identify financial data and intellectual property as high-value assets, and recognize that these assets are vulnerable to phishing attacks, malware, or insider threats. Based on the assessment, the organization can then focus on implementing controls to mitigate these risks.

Risk Management Frameworks: A Structured Approach

Once risks have been identified, organizations should implement a risk management framework to address and mitigate those risks. Risk management frameworks provide a structured approach to identifying, assessing, responding to, and monitoring risks. Several widely recognized frameworks exist, including the NIST Risk Management Framework (RMF), ISO 31000, and **COBIT** (Control Objectives for Information and Related Technologies).

The NIST Risk Management Framework is one of the most widely adopted frameworks in cybersecurity. It provides a systematic process for managing risks, from the initial risk assessment to the ongoing monitoring of implemented security controls. According to NIST, "the RMF integrates security, privacy, and risk management activities into the system development life cycle, ensuring that cybersecurity is addressed throughout the life of a system."[2] The RMF includes the following key steps: prepare, categorize, select, implement, assess, authorize, and monitor.

Similarly, ISO 31000 is a global standard for risk management that applies to a wide range of industries, not just cybersecurity. It emphasizes a holistic approach to managing risks, encouraging organizations to consider both external and internal factors. According to ISO, "ISO 31000 provides principles and guidelines for developing a risk management framework, ensuring that risk management is integrated into organizational processes at all levels "[3].

Risk Treatment: Mitigating Cybersecurity Threats

Once risks have been identified and assessed, organizations should decide how to treat those risks. There are four primary strategies for risk treatment: acceptance, avoidance, mitigation, and transfer. The choice of strategy depends on the nature of the risk and the organization's risk tolerance.

- Risk acceptance involves acknowledging that a risk exists but deciding not to take any immediate action to mitigate it. This approach is typically used when the cost of mitigating the risk outweighs the potential impact of the risk itself.

- Risk avoidance involves eliminating the source of the risk entirely, such as discontinuing the use of a vulnerable system or service. This strategy may be necessary when the risk poses a significant threat to the organization.

- Risk mitigation involves taking steps to reduce the likelihood or impact of a risk. This is the most common approach in cybersecurity and includes measures such as installing firewalls, conducting regular security audits, and implementing encryption.

- Risk transfer involves shifting the responsibility for managing the risk to a third party, such as through cyber insurance or outsourcing specific security functions to a managed service provider. According to Marsh & McLennan, "cyber insurance is increasingly being used as a risk transfer mechanism, helping organizations manage the financial impact of data breaches and cyberattacks."[4]

Integrate Risk Management into Organizational Culture

For risk management frameworks to be effective, they should be integrated into the organization's overall culture and processes. This means that cybersecurity risks should be treated as a core component of the organization's governance, risk, and compliance (GRC) strategy. A successful cybersecurity risk management framework is not only the responsibility of the IT department but also requires collaboration across all business units, from human resources to finance.

Risk management frameworks emphasize the importance of continuous monitoring and auditing to ensure that risks are managed effectively over time. Cyber threats evolve rapidly, and new vulnerabilities can emerge as systems are updated, expanded, or integrated with third-party services. Regular vulnerability assessments and penetration testing help organizations stay ahead of potential risks by identifying weaknesses before they can be exploited. As Peltier notes, "Cybersecurity is not a one-time event but an ongoing process that requires constant vigilance and adaptation."[5]

Organizations should also foster a culture of risk awareness among employees, ensuring that all staff members understand their role in maintaining cybersecurity. Employee training programs can help individuals recognize potential security threats, such as phishing emails or social engineering attacks, and teach them how to respond appropriately. By embedding cybersecurity awareness into the organizational culture, they can reduce the likelihood of human errors that contribute to security breaches.

The Role of Cybersecurity Frameworks in Compliance

In addition to protecting the organization's assets, risk management frameworks help ensure compliance with industry regulations and standards. Many regulatory bodies, such as the General Data Protection Regulation (GDPR)and the Health Insurance Portability and Accountability Act (HIPAA), require organizations to implement risk assessment and management practices as part of their cybersecurity programs.

For example, GDPR mandates that organizations conduct regular data protection impact assessments (DPIAs) to identify and mitigate risks associated with processing personal data. Similarly, HIPAA requires covered entities to implement security measures that reduce the risk of a breach of protected health information (PHI). Failure to comply with these regulations can result in significant fines and legal liabilities. This makes it essential for organizations to integrate compliance into their risk management strategies.

As noted by Kesan and Hayes, "compliance with cybersecurity regulations not only mitigates legal risks but also demonstrates a commitment to data protection, fostering trust among customers and stakeholders."[6] Regular compliance audits and risk assessments ensure that organizations meet regulatory requirements while maintaining a proactive approach to cybersecurity.

The Future of Cybersecurity Risk Management

As cyber threats continue to evolve, so too should the frameworks and strategies used to manage those risks. Emerging technologies such as artificial intelligence (AI) and machine learning (ML)are increasingly being integrated into risk management frameworks to enhance threat detection and response capabilities. AI-powered systems can analyze vast amounts of data in real time, identifying patterns and anomalies that may indicate a potential cyberattack.

According to PwC, "the integration of AI and ML into risk management frameworks offers organizations the ability to detect and respond to threats more quickly and accurately, reducing the likelihood of a successful attack."[7] As the volume and complexity of cyber threats increase, organizations should adopt more advanced technologies to avoid potential risks.

Conclusion

Risk assessment and management frameworks are essential components of a comprehensive cybersecurity strategy. By identifying potential threats, evaluating vulnerabilities, and implementing appropriate risk treatment strategies, organizations can reduce their exposure to cyberattacks and ensure business continuity. Organizations should remain vigilant update their risk management frameworks regularly to address new challenges. By integrating risk management into the organizational culture and leveraging advanced technologies, companies can build a resilient cybersecurity posture that protects their assets and complies with regulatory requirements.

5.3 Implement Security Policies

In today's digital landscape, cyber threats continue to evolve in complexity and scope, making the development and enforcement of robust security policies essential for organizations. A well-structured approach to implementing security policies helps organizations mitigate risks, comply with regulations, and respond effectively to cyber incidents.

The Importance of Security Policies in Cybersecurity

Security policies serve as the foundation for an organization's cybersecurity framework. They provide clear guidelines on protecting sensitive data, responding to security incidents, and ensuring compliance with industry regulations. According to ISO/IEC 27002, "security policies are essential for defining the rules and procedures that ensure the confidentiality, integrity, and availability of information assets."[1]

A comprehensive security policy outlines the organization's security objectives, roles and responsibilities, and acceptable use of technology. These policies should cover areas such as password management, data protection, access control, incident response, and network security. By providing a clear framework, security policies ensure that employees, contractors, and third-party vendors understand their obligations in maintaining cybersecurity.

For example, many organizations implement password policies that require employees to use complex passwords, change them regularly, and avoid reusing passwords across multiple accounts. Additionally, access control policies dictate who has access to specific data and systems, limiting the risk of unauthorized access. Without clearly defined policies, employees may unintentionally engage in risky behaviors, such as sharing passwords or using unsecured devices to access sensitive information.

Developing Effective Security Policies

Security policies should be tailored to the organization's specific needs, considering factors such as industry, size, and regulatory requirements. The National Institute of Standards and Technology (NIST) recommends a risk approach to policy development, ensuring that policies address the most critical risks faced by the organization.[2] This approach involves conducting a thorough risk assessment to identify the organization's most valuable assets and potential threats to those assets.

Once risks are identified, the organization can develop policies that mitigate those risks. For example, a healthcare organization subject to the Health Insurance Portability and Accountability Act (HIPAA)may implement strict data protection policies to ensure the confidentiality of patient health information. Financial institutions, on the other hand, may focus on implementing robust encryption policies to protect financial transactions.

It is also important to ensure that security policies are aligned with industry standards and regulations. Organizations in regulated industries, such as healthcare, finance, and energy, should comply with specific legal requirements related to data protection and cybersecurity. Compliance with frameworks such as GDPR, PCI DSS, and HIPAA requires organizations to implement specific security measures, such as encryption, data access controls, and breach reporting procedures. As Kesan and Hayes note, "security policies should be regularly updated to reflect changes in regulatory requirements and evolving cyber threats."[3]

Developing a Culture of Cybersecurity

Creating a culture of cybersecurity requires more than just formal training sessions. It involves fostering a mindset where cybersecurity is seen as a shared responsibility across the organization. According to SANS Institute, "Cybersecurity should be integrated into the daily routines and decision-making processes of all employees, from entry-level staff to senior executives."[4] A culture of security awareness encourages employees to remain vigilant about potential threats, adopt secure practices in their work, and report suspicious activities promptly.

To build this culture, organizations should promote continuous learning and open communication about cybersecurity. Regular security briefings, email reminders, and interactive workshops can reinforce key concepts and update employees on emerging threats. Simulated phishing exercises are another effective way to test employees' ability to recognize phishing attempts and respond appropriately. These exercises provide valuable insights into the organization's vulnerabilities and help identify areas where additional training may be needed.

In addition to training employees, organizations should also establish clear channels for reporting security incidents. Employees should feel comfortable reporting potential security breaches or suspicious activities without fear of punishment. Encouraging a "see something, say something" approach fosters a proactive attitude toward healthy cybersecurity and allows organizations to address threats before they escalate.

Security Policies for Remote and Hybrid Workforces

The rise of remote work and hybrid work models has introduced new cybersecurity challenges, requiring organizations to update their security policies to account for employees working outside the traditional office environment. Remote work increases the risk of data breaches and cyberattacks, as employees often use personal devices and home networks to access company systems.

To mitigate these risks, organizations should implement security policies that address the specific challenges of remote work. For example, employees should be required to use virtual private networks (VPNs) when accessing company data from home or public networks. Additionally, organizations should enforce strict device security policies, requiring employees to use antivirus software, firewalls, and encryption on personal devices.

Zero Trust Architecture is another important concept for securing remote and hybrid workforces. Zero Trust assumes that no user or device is trusted by default, regardless of whether they are inside or outside the organization's network. This approach requires continuous verification of user identities and device health before granting access to sensitive data. As Forrester Research notes, "the shift to remote work has accelerated the adoption of Zero Trust models, which offer enhanced protection for organizations with distributed workforces."[5]

Ongoing Evaluation and Improvement of Security Policies

It is essential for organizations to regularly evaluate and update their security policies. Annual security audits, penetration testing, and vulnerability assessments can help organizations identify weaknesses in their defenses and ensure that security policies remain effective. Continuous monitoring and feedback from employees can also provide valuable insights into the effectiveness of training programs and highlight areas for improvement.

Organizations should establish a process for reviewing and revising security policies in response to new threats, regulatory changes, and technological advancements. Bertino and Sandhu emphasize that, "Security policies should be living documents that evolve in response to the dynamic nature of cybersecurity risks."[6] Regular updates to policies and training programs ensure that organizations stay ahead of potential threats and maintain a strong security posture.

Conclusion

The critical components of an effective cybersecurity strategy are implementation of comprehensive security policies and providing ongoing employee training for employees. Security policies establish clear guidelines for protecting data and systems, while employee training ensures that staff members are equipped to recognize and respond to cyber threats. By fostering a culture of cybersecurity awareness and regularly updating policies to address emerging risks, organizations can reduce their vulnerability to cyberattacks and protect their valuable assets.

5.4 Incident Response Planning and Execution

In the complex landscape of cybersecurity, it is no longer a question of whether an organization will face a security incident but when. As cyber threats become more sophisticated, the ability to respond swiftly and effectively to a cyberattack is critical for minimizing damage and ensuring business continuity. Incident response planning and execution are key components of an organization's cybersecurity strategy designed to manage and mitigate the consequences of security breaches. A well-prepared incident response plan (IRP) allows organizations to identify, contain, and remediate cyber incidents efficiently, reducing the impact on operations and reputation.

The Importance of Incident Response Planning

An effective incident response plan outlines the procedures and responsibilities necessary to respond to a wide range of security incidents, including data breaches, malware infections, ransomware attacks, and insider threats. The goal of an IRP is to minimize the impact of the incident, restore normal operations, and prevent future occurrences. As noted by ENISA (European Union Agency for Cybersecurity), "Incident response planning is a critical element of an organization's defense strategy, enabling a coordinated and efficient response to security breaches."[1]

A comprehensive IRP includes several key components: preparation, identification, containment, eradication, recovery, and lessons learned. Each phase of the incident response process ensures that the organization can respond promptly to the incident, contain the damage, and implement measures to prevent future breaches. Without a well-developed IRP, organizations risk prolonged downtime, significant financial losses, and reputational damage after a cyberattack.

Phases of Incident Response

1. **Preparation**: The preparation phase involves the development of policies, procedures, and tools necessary to respond to incidents. This includes assembling an incident response team (IRT), defining roles and responsibilities, and conducting regular training exercises. Preparation is key to ensuring that all stakeholders know how to respond in the event of an attack, minimizing confusion and delays.

According to the SANS Institute, "Incident response readiness depends on both technical tools and the preparedness of staff to follow established procedures under pressure."[2] Preparation also involves ensuring that all critical systems are backed up regularly, that security controls are in place, and that the organization's legal and compliance obligations are well-understood.

2. **Identification**: The identification phase focuses on detecting and determining the nature of the incident. This can involve monitoring network traffic for unusual activity, analyzing security logs, and using intrusion detection systems (IDS) to identify anomalies. The faster an organization can detect an incident, the more quickly it can mitigate the damage. Identification also involves determining whether the incident constitutes a true breach or simply an anomaly.

As NIST notes, "early identification of security incidents is critical to limiting their impact, allowing organizations to take immediate action to contain the threat."[3] Timely identification depends on having robust monitoring systems in place and ensuring that staff are trained to recognize potential security incidents.

3. **Containment**: Once an incident has been identified, the organization should act quickly to contain the damage. The containment phase involves isolating affected systems, disconnecting compromised devices from the network, and preventing further unauthorized access. Containment can be divided into short-term and long-term strategies. Short-term containment aims to limit the immediate impact of the attack, while long-term containment focuses on ensuring that the root cause of the breach is addressed.

A real-world example of effective containment occurred during the 2020 SolarWinds breach, where multiple organizations quickly isolated compromised systems to prevent further data exfiltration.[4] Containment measures can include disabling user accounts, shutting down affected servers, or blocking malicious IP addresses. The speed and effectiveness of containment efforts directly influence the overall impact of the attack.

4. **Eradication**: After containing the incident, the next step is to eradicate the root cause of the breach. This involves removing malware, closing security vulnerabilities, and ensuring that unauthorized access points are fully eliminated. Forensic analysis is often conducted during this phase to understand how the breach occurred and what vulnerabilities were exploited.

Eradication is essential for preventing attackers from re-entering the network. According to Verizon's 2021 Data Breach Investigations

Report, "failure to fully eradicate the root cause of a breach can lead to recurring incidents, prolonging the recovery process and increasing costs."[5] Organizations should ensure that all compromised systems are thoroughly cleansed and that security patches are applied before resuming normal operations.

5. **Recovery**: The recovery phase focuses on restoring affected systems and data to their pre-incident state. This includes restoring backups, patching vulnerabilities, and monitoring systems for any signs of residual compromise. Recovery efforts should be carefully managed to avoid reintroducing the threat to the network.

Recovery time is critical, as prolonged downtime can lead to significant financial losses and damage to customer trust. A successful recovery plan involves not only restoring systems but also ensuring that the organization's operations are back to full functionality. Disaster recovery plans are often invoked during this phase to ensure business continuity. According to IBM's Cost of a Data Breach Report, "the average time to recover from a data breach is 287 days, making efficient recovery strategies essential for minimizing financial impact."[6]

6. **Lessons Learned**: After the incident has been resolved, the final phase of incident response involves conducting a post-incident review to analyze what went wrong and how future incidents can be prevented. The lessons learned phase is crucial for improving the organization's security posture and strengthening defenses against future attacks.

According to Bodeau and Graubart, "the lessons learned phase allows organizations to identify gaps in their incident response processes and implement new controls to prevent similar breaches from occurring in the future."[7] This phase may involve updating security policies, conducting additional employee training, or investing in new security tools.

Incident Response Team (IRT) and Roles

An effective incident response plan requires a well-coordinated incident response team (IRT) composed of individuals with specific roles and responsibilities. The IRT should include representatives from IT, legal, human resources, public relations, and senior management. Each member of the team plays a critical role in managing the organization's response to a security incident.

For example, the IT team is responsible for identifying, containing, and eradicating the threat, while the legal team ensures that the organization complies with data breach notification laws and regulatory requirements. The public relations team manages communications with customers and the media to protect the organization's reputation. As Crowe and Riley point out, "a well-defined incident response team ensures that all aspects of the incident, from technical resolution to public communication, are managed effectively."[8]

The Role of Cyber Insurance in Incident Response

In recent years, cyber insurance has become an important part of incident response planning. Cyber insurance policies can help organizations manage the financial impact of a cyberattack, covering costs related to data breaches, ransomware payments, legal fees, and business interruption. According to Marsh McLennan, "Cyber insurance is a critical tool for mitigating the financial impact of cyber incidents, particularly for small and medium-sized enterprises that may not have the resources to recover from a major attack."[9]

However, organizations should ensure that they meet the security requirements set by their cyber insurance policies, as failure to comply with these requirements can result in denied claims. For example, some policies require organizations to implement multi-factor authentication (MFA) and conduct regular security audits to qualify for coverage.

Conclusion

Incident response planning and execution are essential for managing the increasing complexity of cyber threats. A well-prepared incident response plan allows organizations to respond quickly and effectively to cyber incidents, minimizing damage and ensuring business continuity. By following the key phases of incident response, organizations can strengthen their defenses and reduce the long-term impact of security breaches. Organizations should remain vigilant and continuously update their incident response plans to address emerging risks.

5.5 Business Continuity and Disaster Recovery

In the face of increasing cyber threats, business continuity and disaster recovery (BC/DR) strategies have become essential components of an organization's cybersecurity framework. These strategies ensure that businesses can maintain operations and quickly recover after disruptions caused by cyberattacks, natural disasters, or other unforeseen events. A well-developed BC/DR plan minimizes downtime, protects critical assets, and ensures that organizations can continue delivering services even during a crisis.

The Importance of Business Continuity in Cybersecurity

Business continuity (BC)refers to the processes and procedures that organizations implement to ensure that critical operations can continue during and after a disruptive event. Cyber incidents, such as ransomware attacks or data breaches, can cause significant disruptions to business operations, leading to lost revenue, reputational damage, and regulatory penalties. According to IBM's Cost of a Data Breach Report, "the average cost of downtime due to a cyber incident is estimated to be $4.24 million."[1] A robust business continuity plan helps organizations minimize these losses by ensuring that key functions can continue with minimal interruption.

A key element of business continuity is the identification of critical business functions and key dependencies. These include essential operations that should be maintained during a crisis, such as customer services, supply chain management, and IT infrastructure. By identifying these critical components, organizations can prioritize their recovery efforts and ensure that resources are allocated to the most important areas. Additionally, BC plans should include clear communication protocols to ensure that employees, customers, and stakeholders are informed about the status of operations during a disruption.

Disaster Recovery

Disaster recovery (DR)is the process of restoring systems, data, and infrastructure to normal functioning following a disruptive event. While business continuity focuses on maintaining operations during a crisis, disaster recovery focuses on returning systems to their pre-incident state. The primary goal of disaster recovery is to restore data integrity and system functionality as quickly as possible, minimizing the impact on the organization's operations and reputation.

A critical component of disaster recovery is the establishment of recovery time objectives (RTOs) and recovery point objectives (RPOs). RTO refers to the maximum acceptable amount of time that systems can be down before the business experiences unacceptable losses, while RPO refers to the maximum amount of data that can be lost during an incident without causing significant harm to the business. For example, a financial services company may have an RTO of two hours and an RPO of 15 minutes, meaning that systems should be restored within two hours, and no more than 15 minutes of data can be lost.

According to Gartner, "disaster recovery plans that clearly define RTOs and RPOs enable organizations to prioritize recovery efforts and allocate resources efficiently."[2] Effective disaster recovery plans also include backup strategies that ensure critical data is stored in secure, offsite locations. Cloud-based backups and disaster recovery as a service (DRaaS) are increasingly popular options for organizations looking to enhance their resilience in the face of cyberattacks and other disasters.

Developing a Business Continuity and Disaster Recovery Plan

Creating a comprehensive BC/DR plan involves several key steps, beginning with a business impact analysis (BIA). The BIA identifies the potential impact of different types of disruptions on the organization's operations, finances, and reputation. This analysis helps organizations prioritize their recovery efforts by identifying which systems and processes are most critical to business continuity. As noted by The Business Continuity Institute, "a thorough business impact analysis provides the foundation for a successful BC/DR plan by highlighting the areas that require the most attention during a crisis."[3]

The next step in developing a BC/DR plan is to establish recovery strategies for maintaining critical operations and restoring systems. These strategies may include redundant systems, offsite data storage, and failover procedures, which allow organizations to switch to backup systems in the event of a failure. For example, a healthcare organization may implement a failover system that automatically redirects data processing to a secondary data center if the primary center experiences an outage.

Another important element of BC/DR planning is the development of communication plans that outline how the organization will communicate with employees, customers, suppliers, and other stakeholders during and after an incident. Clear and transparent communication is essential for maintaining trust and minimizing the reputational damage associated with cyberattacks and other disruptions as Paul and Murry explain, "Organizations that communicate effectively during a crisis are more likely to retain customer loyalty and recover more quickly from the incident."[4]

Testing and Updating BC/DR Plans

A BC/DR plan is only effective if it is regularly tested and updated to reflect changes in the organization's operations, technology, and threat landscape. Tabletop exercises and full-scale simulations allow organizations to test their plans under realistic conditions, identifying potential weaknesses and areas for improvement. According to SANS Institute, "regular testing of BC/DR plans is essential for ensuring that employees understand their roles and responsibilities during an incident and that the plan is capable of delivering the expected outcomes."[5]

Testing also provides valuable insights into how well the organization's backup and recovery systems function during an actual disruption. For example, a company may discover during testing that its data recovery times are slower than expected, prompting an upgrade to its backup infrastructure. Additionally, BC/DR plans should be updated to account for changes in business processes, such as the introduction of new technologies, changes in the supply chain, or the adoption of cloud services. Continuous improvement is critical for ensuring that the organization remains resilient in the face of evolving cyber threats.

The Role of Cloud Computing in Business Continuity and Disaster Recovery

The rise of cloud computing has transformed the way organizations approach business continuity and disaster recovery. Cloud-based solutions offer increased flexibility, scalability, and cost-effectiveness compared to traditional on-premise systems. With cloud-based backups and DRaaS, organizations can store critical data in multiple locations, reducing the risk of data loss due to physical damage or cyberattacks. Cloud-based systems also allow for faster recovery times, as data can be restored from the cloud without the need for physical hardware.

For example, after a ransomware attack, an organization that uses cloud-based backups can quickly restore its data from the cloud, bypassing the need to pay a ransom. As IDC notes, "the adoption of cloud-based disaster recovery solutions has significantly reduced recovery times for organizations, allowing them to resume operations more quickly after an incident."[6] Cloud-based BC/DR solutions are particularly beneficial for small and medium-sized businesses, which may lack the resources to maintain extensive on-premise backup systems.

Regulatory Compliance and BC/DR

In many industries, business continuity and disaster recovery plans are not just a best practice but a regulatory requirement. Regulatory frameworks such as the General Data Protection Regulation (GDPR)and the Health Insurance Portability and Accountability Act (HIPAA) mandate that organizations implement measures to ensure the protection and recovery of sensitive data in the event of a disruption. Failure to comply with these regulations can result in significant fines and reputational damage.

For example, GDPR requires organizations that process personal data to have appropriate technical and organizational measures in place to ensure the availability and resilience of processing systems. This includes the implementation of regular testing and disaster recovery capabilities to safeguard data in the event of a physical or technical incident. As Kesan and Hayes explain, "organizations should integrate regulatory requirements into their BC/DR plans to ensure compliance and avoid costly penalties."[7]

Conclusion

Business continuity and disaster recovery are essential components of an organization's cybersecurity strategy. It ensures that critical operations can continue and systems can be restored after a disruptive event. A well-developed BC/DR plan minimizes downtime, protects valuable assets, and helps organizations recover more quickly from cyberattacks and other incidents. By conducting regular testing, leveraging cloud-based solutions, and ensuring compliance with regulatory requirements, organizations can enhance their resilience in the face of an increasingly complex threat landscape.

5.6 Third-Party Risk Management and Supply Chain Security

In today's interconnected business environment, organizations often rely on third-party vendors and supply chain partners to deliver goods, services, and critical IT infrastructure. However, this reliance exposes organizations to new cybersecurity risks, as vulnerabilities within a third party can lead to data breaches, operational disruptions, and significant reputational damage. Implementing third-party risk management (TPRM) and supply chain security measures are essential to mitigating these risks and ensuring that external partners maintain the same high standards of cybersecurity as the organization itself.

The Growing Threat of Supply Chain Attacks

Cybercriminals are increasingly targeting third-party vendors and supply chains as a way to gain access to the networks of larger organizations. These attacks exploit weaknesses in the security practices of third-party service providers, allowing attackers to infiltrate organizations through indirect means. As noted by KPMG, "supply chain attacks have increased by 78% in the last year, as cybercriminals seek to exploit the weakest links in complex, interconnected networks."[1]

A notable example of a supply chain attack is the 2020 SolarWinds breach, where attackers compromised the software provider SolarWinds to infiltrate multiple government agencies and corporations. By inserting malware into a routine software update, attackers were able to gain access to the networks of SolarWinds' clients, highlighting the dangers of trusting third-party vendors without rigorous security controls[2]. This breach underscored the importance of closely monitoring third-party vendors and regularly assessing their security practices to prevent similar attacks.

Third-Party Risk Management (TPRM): A Holistic Approach

Third-party risk management involves identifying, assessing, and mitigating the cybersecurity risks posed by external vendors, contractors, and service providers. Effective TPRM requires a holistic approach that includes vendor risk assessments, contractual agreements, continuous monitoring, and incident response planning. As organizations rely more heavily on third parties for critical business functions, TPRM becomes an integral part of maintaining robust cybersecurity.

The first step in TPRM is conducting a thorough vendor risk assessment before entering into a contract with a third party. This assessment should evaluate the vendor's cybersecurity posture, including their security policies, data protection practices, and compliance with industry standards. According to Forrester Research, "Organizations should assess the security risks of their third-party vendors to ensure that they meet the same cybersecurity requirements as the organization itself."[3] This risk assessment may include reviewing the vendor's security certifications (such as ISO/IEC 27001), conducting penetration testing, and verifying that they have adequate controls in place to protect sensitive data.

Contractual obligations also play a crucial role in TPRM. Contracts should include specific cybersecurity requirements, such as data protection measures, incident reporting procedures, and provisions for auditing the vendor's security practices. By clearly defining these expectations in the contract, organizations can ensure that third-party vendors adhere to established security standards. For example, a contract with a cloud service provider might include clauses that require the vendor to encrypt data at rest and in transit, conduct regular security audits, and notify the organization of any data breaches within a specified timeframe.

Continuous Monitoring of Third-Party Risks

Once a vendor relationship is established, organizations should engage in continuous monitoring to identify any emerging risks or vulnerabilities. Continuous monitoring involves regularly assessing the vendor's security practices, conducting audits, and tracking any changes in their risk profile. This ensures that the vendor maintains a high level of cybersecurity throughout the duration of the relationship.

A real-world example of the importance of continuous monitoring is the 2013 Target breach, where attackers exploited a vulnerability in the security practices of a third-party HVAC contractor. The contractor had access to Target's network, and when their credentials were compromised, the attackers were able to move laterally within Target's systems, ultimately stealing the credit card information of 40 million customers.[4] Had Target implemented continuous monitoring of the contractor's security practices, the breach might have been detected earlier and the damage mitigated.

Supply Chain Security: Managing Risks in the Supply Chain

Supply chain security focuses on protecting the integrity and security of products, services, and information as they move through the supply chain. As organizations become more global and interconnected, the complexity of their supply chains increases, leading to greater exposure to cyber risks. Effective supply chain security requires organizations to identify and address vulnerabilities at each stage of the supply chain, from sourcing raw materials to delivering finished products.

One of the key challenges in supply chain security is ensuring the authenticity and integrity of software and hardware components. Cybercriminals may target suppliers of software and hardware products, inserting backdoors or malware into products before they reach the customer. As PwC notes, "supply chain attacks on software and hardware products pose a significant risk, as compromised components can introduce vulnerabilities that are difficult to detect and mitigate."[5] To manage these risks, organizations should implement vendor vetting procedures, require code signing for software products, and conduct thorough testing of hardware components to verify their security.

Another critical aspect of supply chain security is ensuring that third-party vendors have robust incident response plans in place. In the event of a cybersecurity incident, the vendor should be able to quickly contain and mitigate the damage while also coordinating with the organization to ensure that the supply chain is not disrupted. As Sinha et al. explain, "Incident response coordination between organizations and their supply chain partners is essential for minimizing the impact of a cyberattack on critical operations." [6]

The Role of Cybersecurity Standards and Regulations

Regulatory frameworks and industry standards play an important role in third-party risk management and supply chain security. Compliance with these standards ensures that organizations and their third-party vendors implement adequate security controls to protect sensitive data and maintain the integrity of the supply chain. Some of the most widely recognized standards include ISO/IEC 27001 for information security management and NIST's Cybersecurity Framework, which provides guidelines for managing cybersecurity risks in critical infrastructure.

In addition to industry standards, regulatory frameworks such as the General Data Protection Regulation (GDPR) and the California Consumer Privacy Act (CCPA) impose specific requirements for third-party vendors. These regulations require organizations to ensure that their vendors implement appropriate security measures to protect personal data and to provide transparency regarding the handling of that data. Failure to comply with these regulations can result in significant financial penalties and reputational damage.

According to Gartner, "ensuring third-party compliance with cybersecurity standards and regulations is critical for reducing the risk of supply chain attacks and maintaining customer trust."[7] Organizations should regularly audit their third-party vendors to verify that they comply with relevant regulations and that they continue to meet the security requirements outlined in their contracts.

Emerging Trends in Third-Party Risk Management

As cyber threats evolve, so too should the strategies for managing third-party risks and securing supply chains. One emerging trend is the use of artificial intelligence (AI) and machine learning (ML) to improve third-party risk assessments and monitoring. AI-powered tools can analyze vast amounts of data to identify potential vulnerabilities in third-party systems and detect anomalous behavior that may indicate a security breach. These tools provide organizations with real-time insights into the security posture of their vendors, allowing for more proactive risk management.

Additionally, Blockchain technology is being explored as a means of securing supply chains by providing a tamper-proof record of transactions and product movements. Blockchain's distributed ledger technology can help ensure the authenticity and integrity of products as they move through the supply chain, reducing the risk of tampering or counterfeiting.

As Meidan et al. argue, "the integration of AI and Blockchain into third-party risk management and supply chain security offers organizations new tools for enhancing visibility and resilience in their cybersecurity strategies."[8]

Conclusion

In an increasingly interconnected world, third-party risk management and supply chain security are critical to protecting organizational assets and maintaining business continuity. By conducting thorough vendor risk assessments, implementing continuous monitoring, and ensuring compliance with industry standards, organizations can mitigate the cybersecurity risks posed by third-party vendors. Organizations should adopt innovative technologies such as AI and Blockchain to strengthen their third-party risk management strategies and secure their supply chains. Effective management of third-party risks not only protects organizations from cyberattacks but also fosters trust and resilience across the entire supply chain.

Test Your Knowledge on Organizational Cybersecurity

Q1: What are the key components of a comprehensive cybersecurity strategy for organizations?

Q2: How does risk assessment contribute to organizational cybersecurity?

Q3: What role do security policies play in organizational cybersecurity?

Q4: How can organizations effectively train employees to recognize and prevent cyber threats?

Q5: Why is incident response planning critical for organizations?

Q6: What are the essential elements of a well-designed incident response plan?

Q7: How does business continuity planning relate to cybersecurity?

Q8: What are the benefits of third-party risk management in cybersecurity?

Q9: How can organizations implement privileged access management (PAM)?

Q10: What role does employee awareness play in preventing insider threats?

Q11: How do organizations measure the effectiveness of their cybersecurity strategies?

Q12: Why is security governance important in organizational cybersecurity?

Q13: How do organizations ensure compliance with cybersecurity regulations?

Q14: What is the role of cybersecurity frameworks, such as NIST, in organizational defense?

Q15: How can organizations foster a culture of security within the workplace?

ANSWERS

Q1: A: A comprehensive strategy includes risk assessment, incident response planning, employee training, security policies, and continuous monitoring.

Q2: A: Risk assessment identifies potential vulnerabilities, evaluates the impact of threats, and helps organizations prioritize resources to address the most critical risks.

Q3: A: Security policies set guidelines for employees, contractors, and vendors on how to handle data, access systems, and respond to security incidents.

Q4: A: Effective training includes educating employees about phishing, password hygiene, social engineering, and data protection, with regular updates and testing.

Q5: A: Incident response planning ensures that organizations can quickly detect, contain, and mitigate security breaches, minimizing damage and downtime.

Q6: A: Essential elements include detection protocols, communication plans, mitigation procedures, forensic analysis, and recovery processes.

Q7: A: Business continuity planning ensures that an organization can continue operating after a cyberattack by having backups, recovery strategies, and contingency plans in place.

Q8: A: Third-party risk management ensures that vendors and partners meet security standards, reducing the risk of breaches through third-party connections.

Q9: A: Organizations can implement PAM by restricting access to sensitive systems, monitoring user activities, and enforcing the principle of least privilege.

Q10: A: Employee awareness is crucial, as insider threats often result from negligence or malicious actions. Regular training and monitoring can help prevent these risks.

Q11: A: Effectiveness is measured through metrics like mean time to detect(MTTD), mean time to respond (MTTR), incident response success, and system uptime.

Q12: A: Security governance aligns cybersecurity efforts with business objectives, ensuring that security practices are integrated into organizational decision-making and risk management.

Q13: A: Compliance is ensured by staying informed about relevant laws, such as GDPR or HIPAA, conducting regular audits, and implementing security controls that meet regulatory requirements.

Q14: A: Cybersecurity frameworks provide a structured approach to managing risks, defining best practices, and aligning security efforts with recognized standards.

Q15: A: Organizations can foster a security culture by promoting security awareness, encouraging collaboration between departments, and rewarding compliance with security policies.

CHAPTER 6

CYBERSECURITY IN KEY INDUSTRIES

Note: All references in this chapter are on page 401 - 402

I n this chapter we will investigate how, depending on their operations, legal requirements, and digital infrastructure, many businesses confront particular cybersecurity issues. Because the financial services sector handles so much sensitive data and conducts safe transactions and fraud protection is so important, so fraudsters target this sector most especially.

While safeguarding patient data and medical equipment from cyberattacks, healthcare has to abide by rigorous rules, including HIPAA. Embracing smart grids and linked devices at an accelerating speed, the energy and utility sectors should protect important infrastructure against cyberattacks capable of causing major disruptions. Likewise, continuous threats to the retail and e-commerce industries are payment card theft, data breaches, and attacks on consumer confidence.

Examining every business using case studies and real-world examples reveals the several types of attacks and legal restrictions they come across. By the end of this chapter, you will understand the need to protect vital infrastructure as well as the particular cybersecurity strategies needed for every one of these sectors. The chapter will also show how these industries may welcome innovative technologies while lowering the specific risks they run into.

6.1 Financial Services: Protecting Sensitive Transactions and Data

The financial services sector is one of the most highly regulated and targeted industries in the world, making cybersecurity a top priority. Financial institutions handle vast amounts of sensitive data, including personal information, financial transactions, and proprietary business information. Given the critical nature of this data, protecting it from cyber threats is essential for maintaining customer trust, complying with regulatory requirements, and preventing financial losses. Cyberattacks on the financial services sector have become increasingly sophisticated, requiring institutions to adopt robust security measures to safeguard sensitive transactions and data.

The Rising Threat of Cyberattacks on Financial Institutions

Cybercriminals are increasingly targeting financial institutions due to the high value of the data they hold and the potential financial gain from successful attacks. Common types of attacks include phishing, ransomware, account takeover, and data breaches. As noted by Accenture's 2021 Cybercrime Report, "the financial services sector is 300% more likely to experience a cyberattack than other industries, highlighting the critical need for strong cybersecurity measures."[1]

One of the most concerning trends is the rise of ransomware attacks, where cybercriminals encrypt an organization's data and demand a ransom to restore access. These attacks can cause significant operational disruptions, as financial institutions are unable to process transactions or access critical data. In 2020, the CNA Financial ransomware attack caused major disruptions, leading to a $40 million ransom payment.[2] This incident underscores the importance of having comprehensive disaster recovery and incident response plans in place to quickly recover from such attacks.

Protect Sensitive Financial Data

The protection of sensitive financial data is at the core of cybersecurity efforts in the financial services sector. Financial institutions store a wealth of personal and financial information about their customers, including credit card numbers, bank account details, and Social Security numbers. If this data is compromised, it can lead to identity theft, fraud, and significant financial losses for both the institution and its customers.

To protect sensitive data, financial institutions should implement a combination of technical controls, regulatory compliance, and best practices. Data encryption is one of the most effective ways to protect data, ensuring that even if information is intercepted or stolen, it remains unreadable to unauthorized parties. End-to-end encryption is commonly used in online banking and payment processing to secure transactions from the point of initiation to the final destination.

Tokenization is another method used to protect financial data by replacing sensitive information, such as credit card numbers, with a unique token that can only be decoded by authorized systems. According to PCI Security Standards Council, "tokenization is a powerful tool for reducing the risk of data breaches, as it ensures that sensitive information is not stored or transmitted in its original form."[3]

Multi-factor authentication (MFA)is also critical for securing access to financial data. MFA requires users to provide two or more forms of verification before accessing an account, such as a password and a fingerprint or a one-time code sent to a mobile device. As Verizon's 2021 Data Breach Investigations Report highlights, "accounts protected by multi-factor authentication are significantly less likely to be compromised, as attackers should bypass multiple layers of security."[4]

Regulatory Compliance in Financial Services

The financial services industry is subject to a wide range of regulatory requirements designed to protect sensitive data and ensure the integrity of financial transactions. These regulations vary by region and include frameworks such as GDPR (General Data Protection Regulation), PCI DSS (Payment Card Industry Data Security Standard), SOX (Sarbanes-Oxley Act), and GLBA (Gramm-Leach-Bliley Act). Compliance with these regulations is mandatory for financial institutions, and failure to comply can result in significant fines and legal liabilities.

PCI DSS is one of the most widely recognized standards for protecting cardholder data. It requires financial institutions to implement strict security controls, including encryption, access controls, and regular security audits. The standard applies to any organization that processes, stores, or transmits credit card information. According to the PCI Security Standards Council, "compliance with PCI DSS is essential for preventing data breaches and ensuring the security of payment card transactions."[5]

Similarly, GDPR imposes strict requirements for protecting personal data and gives individuals more control over how their data is collected and used. Financial institutions that process personal data from EU citizens should comply with GDPR's requirements, including the obligation to report data breaches within 72 hours and implement measures to protect data privacy. As noted by Kesan and Hayes, "GDPR compliance is not only a legal requirement but also a competitive advantage, as customers increasingly prioritize data protection and privacy."[6]

Prevent Fraud and Secure Financial Transactions

In addition to protecting sensitive data, financial institutions should also focus on and preventing fraud and securing financial transactions. Fraud prevention involves detecting and stopping fraudulent transactions before they occur, while transaction security ensures that legitimate transactions are processed securely. Financial institutions use a combination of fraud detection systems, transaction monitoring, and behavioral analysis to identify suspicious activity and prevent fraud.

Real-time transaction monitoring is one of the most effective tools for preventing fraud. By analyzing transaction patterns in real time, financial institutions can detect anomalies, such as unusual spending patterns or transactions from unfamiliar locations, and flag them for further investigation. For example, if a customer's account is accessed from a foreign country shortly after a transaction in their home country, the system may trigger a fraud alert.

Machine learning (LM) and artificial intelligence (AI) are increasingly being used to enhance fraud detection and transaction monitoring. AI-powered systems can analyze vast amounts of data to identify patterns and detect potential fraud more accurately than traditional methods. As noted by PwC, "the integration of AI and machine learning into fraud detection systems allows financial institutions to improve the accuracy of fraud detection and reduce false positives."[7]

Incident Response and Cyber Resilience

Despite the best security measures, cyberattacks on financial institutions are inevitable. Therefore, having a robust incident response plan is critical for minimizing the impact of an attack and ensuring a swift recovery. An incident response plan outlines the steps the organization will take in the event of a cyberattack, from identifying the breach to containing the damage and recovering operations.

According to SANS Institute, "effective incident response planning requires collaboration between IT, legal, and public relations teams to manage the technical, legal, and reputational aspects of the breach."[8] This ensures that financial institutions can respond quickly to cyber incidents, minimize downtime, and maintain customer trust.

Cyber resilience is another key component of cybersecurity in the financial services sector. Cyber resilience refers to an organization's ability to withstand, respond to, and recover from cyberattacks. This includes implementing disaster recovery plans, data backups, and business continuity strategies to ensure that operations can continue even in the event of a major cyber incident. As Deloitte points out, "cyber resilience is essential for financial institutions to maintain trust and confidence in the financial system, particularly in the face of increasing cyber threats." [9]

Conclusion

The financial services sector faces unique cybersecurity challenges due to the sensitive nature of the data it handles and the high value of its transactions. Protecting this data requires a combination of robust technical controls, regulatory compliance, fraud prevention measures, and incident response planning. By implementing strong encryption, multi-factor authentication, and real-time transaction monitoring, financial institutions can reduce the risk of cyberattacks and protect sensitive financial data. Additionally, compliance with regulatory frameworks such as PCI DSS and GDPR is essential for maintaining the security and privacy of customer data. Financial institutions should remain vigilant and invest in advanced technologies to strengthen their defenses and ensure the security of the global financial system.

6.2 Healthcare: Secure Patient Information and Medical Devices

In the digital age, the healthcare industry faces significant cybersecurity challenges as it increasingly adopts digital tools and connected devices. Securing patient information and medical devices is critical for maintaining patient trust, ensuring regulatory compliance, and safeguarding lives. Healthcare organizations store vast amounts of sensitive data, including electronic health records (EHRs), medical histories, financial information, and insurance details. Additionally, the growing use of Internet of Medical Things (IoMT) devices, such as pacemakers and insulin pumps, has introduced new vulnerabilities that cybercriminals can exploit. A robust approach to cybersecurity is therefore, essential to protect both patient data and medical devices from cyber threats.

The Vulnerability of Patient Data

Healthcare organizations are prime targets for cyberattacks due to the value of the data they store. Patient health information (PHI)is highly sought after on the black market, as it can be used for identity theft, insurance fraud, or to create counterfeit medical documents. According to a report by Protenus, "the average cost of a healthcare data breach in 2022 was $10.1 million, the highest of any industry."[1] The impact of a data breach in healthcare extends beyond financial losses, as compromised patient data can result in harm to patients' privacy and medical care.

One of the most common types of cyberattacks in the healthcare industry is ransomware. In 2020, a ransomware attack on UHS Hospitals led to widespread disruption, forcing the hospital network to cancel surgeries, reroute patients, and revert to paper records for several days.[2] This incident highlighted the critical need for healthcare organizations to implement robust backup systems and disaster recovery plans to quickly recover from such attacks.

Protect Electronic Health Records (EHRs)

Electronic health records (EHRs) are one of the most valuable assets in the healthcare industry, containing detailed patient histories, diagnoses, treatment plans, and laboratory results. Securing EHRs is essential for maintaining patient privacy and ensuring that healthcare providers have accurate information to deliver care. The Health Insurance Portability and Accountability Act (HIPAA) sets strict guidelines for protecting patient data and requires healthcare organizations to implement safeguards, including encryption, access controls, and regular audits.

Data encryption is one of the most effective ways to protect EHRs. This ensures that patient information remains secure even if it is intercepted by unauthorized users. Encryption transforms data into unreadable code, which can only be decrypted by authorized personnel with the correct key. Many healthcare organizations use end-to-end encryption to protect data as it is transmitted between different systems, such as between a doctor's office and a pharmacy.

In addition to encryption, healthcare organizations should implement access control mechanisms to restrict who can view or edit patient records. According to Verizon's Data Breach Investigations Report, "insider threats account for a significant percentage of healthcare data breaches, making it essential to limit access to sensitive data based on job roles and responsibilities."[3] This can be achieved through role-based access control (RBAC), where employees are only given access to the information they need to perform their duties.

The Rise of Internet of Medical Things (IoMT) Devices

The adoption of Internet of Medical Things (IoMT) devices has transformed patient care by enabling real-time monitoring, remote diagnostics, and automated treatment. These devices, which include wearable health trackers, smart pacemakers, and insulin pumps, are often connected to hospital networks and can communicate with other medical systems. While IoMT devices offer significant benefits, they also introduce new cybersecurity risks.

IoMT devices are vulnerable to cyberattacks due to their connection to the internet and lack of built-in security features in some of these devices. In some cases, cybercriminals can exploit these devices to gain access to hospital networks or interfere with the functionality of the devices themselves. For example, a 2017 study by the U.S. Food and Drug Administration (FDA) found that certain pacemakers could be remotely hacked, potentially allowing attackers to alter the device's settings and endanger the patient's life.[4]

To protect IoMT devices from cyber threats, healthcare organizations should implement a combination of network segmentation, regular software updates, and device monitoring. Network segmentation involves isolating IoMT devices on separate networks from the main hospital system. This limits the potential damage in the event of an attack. Additionally, manufacturers of medical devices should regularly issue security patches to address vulnerabilities, and healthcare providers should ensure that these updates are applied promptly.

Regulatory Compliance in Healthcare Cybersecurity

Compliance with regulatory frameworks is a key aspect of healthcare cybersecurity. The Health Insurance Portability and Accountability Act (HIPAA) is the primary regulation governing the protection of patient data in the United States. HIPAA requires healthcare organizations to implement both physical and technical safeguards to protect patient information, including encryption, access controls, and security audits. Failure to comply with HIPAA regulations can result in significant fines and reputational damage.

In addition to HIPAA, healthcare organizations should also comply with the General Data Protection Regulation (GDPR) if they handle personal data from EU citizens. GDPR imposes strict requirements for data protection and breach notification, including the obligation to report data breaches within 72 hours. According to Kesan and Hayes, "Healthcare organizations that operate globally should ensure compliance with both HIPAA and GDPR to avoid legal liabilities and protect patient privacy."[5]

The Role of Multi-Factor Authentication (MFA)

Multi-factor authentication (MFA) is another critical tool for securing healthcare systems and protecting patient information. MFA requires users to provide two or more forms of verification before gaining access to sensitive systems. This additional layer of security makes it more difficult for cybercriminals to gain unauthorized access, even if a password is compromised.

Many healthcare organizations have adopted MFA as part of their cybersecurity strategy to protect EHRs and other sensitive data. According to Deloitte, "the use of multi-factor authentication in healthcare has significantly reduced the number of unauthorized access incidents, as attackers should overcome multiple security barriers to gain entry."[6] MFA is particularly important in environments where multiple employees share access to the same system, such as hospitals or clinics.

Incident Response and Cyber Resilience in Healthcare

Given the high stakes of cyberattacks in healthcare, having a well-developed incident response plan is critical for minimizing the impact of a breach and ensuring patient safety. An incident response plan outlines the steps that healthcare organizations will take in the event of a cyberattack, including how to identify the breach, contain the damage, and restore normal operations. Cyber resilience refers to the organization's ability to continue providing care and maintaining operations even in the face of a cyberattack.

One of the key challenges in healthcare incident response is ensuring that patient care is not disrupted. In the case of a ransomware attack, for example, healthcare providers may need to revert to manual processes, such as paper records, to continue delivering care while the IT team works to restore systems. According to SANS Institute, "effective incident response in healthcare requires coordination between IT, clinical staff, and management to minimize the impact on patient care."[7]

The Future of Healthcare Cybersecurity

As healthcare organizations continue to adopt digital technologies, the need for advanced cybersecurity measures will only grow. Artificial intelligence (AI) and machine learning (ML) are emerging as powerful tools for detecting and responding to cyber threats in real time. AI-powered systems have the capabilities to analyze vast amounts of data to identify patterns and detect anomalies that may indicate a cyberattack, allowing healthcare organizations to respond more quickly and effectively.

In addition to AI, the use of Blockchain technology in healthcare is being explored as a way to improve the security and privacy of patient data. Blockchain's distributed ledger technology provides a secure and transparent way to store and share medical records, reducing the risk of data breaches. According to PwC, "Blockchain has the potential to revolutionize healthcare cybersecurity by providing a tamper-proof method for managing patient data and ensuring the integrity of medical records."[8]

Conclusion

Securing patient information and medical devices is a critical challenge for healthcare organizations in the digital age. The increasing adoption of EHRs and IoMT devices has introduced new vulnerabilities, making it critical for healthcare providers to implement robust cybersecurity measures. By adopting encryption, access controls, multi-factor authentication, and incident response plans, healthcare organizations can protect patient data and ensure that medical devices remain secure. There is an urgent need for healthcare providers to remain vigilant and invest in advanced technologies to strengthen their defenses and maintain patient trust as cyber threats continue to evolve.

6.3 Energy and Utilities: Safeguarding Critical Infrastructure

The energy and utilities sector plays a vital role in powering the world's economies, supporting essential services, and maintaining public safety. As critical infrastructure, energy systems—including power grids, water systems, and oil pipelines—are prime targets for cyberattacks. Safeguarding this infrastructure is paramount to ensuring national security, economic stability, and the well-being of citizens. Cybersecurity in the energy and utilities sector is focused on protecting systems from malicious actors seeking to disrupt services, steal data, or cause physical harm. The increasing digitization of energy systems has expanded the attack surface, making the need for robust cybersecurity measures more pressing than ever.

The Growing Threat of Cyberattacks on Critical Infrastructure

Cyberattacks on energy and utility systems are growing in frequency and sophistication, with cybercriminals, nation-state actors, and hacktivists increasingly targeting critical infrastructure. These attacks can have severe consequences, ranging from prolonged power outages and disrupted water supplies to widespread economic damage. According to a report by Dragos, "the energy sector faces a significant threat from cyberattacks, with industrial control systems (ICS) being a primary target for disrupting operations."[1] ICS are the backbone of critical infrastructure, controlling key processes such as electricity distribution and water treatment. A successful attack on these systems could lead to cascading failures across other sectors, including healthcare, transportation, and communications.

A high-profile example of such an attack is the 2015 Ukraine power grid hack, where cybercriminals infiltrated the country's power grid, causing widespread blackouts that affected over 230,000 people [2]. The attackers used spear-phishing emails to gain access to the control systems of the power grid, highlighting the vulnerability of energy infrastructure to human error and social engineering attacks. This incident underscores the critical need for energy and utility companies to enhance their cybersecurity defenses.

Protect Industrial Control Systems (ICS)

Industrial control systems (ICS) are integral to the operation of critical infrastructure in the energy and utilities sector. These systems include supervisory control and data acquisition (SCADA) systems, distributed control systems (DCS), and programmable logic controllers (PLCs), which monitor and control physical processes such as energy generation and water treatment. Protecting these systems is essential to ensuring continuous operation of critical infrastructure.

One of the primary challenges in securing ICS is that many of these systems were not designed with cybersecurity in mind. Traditionally, ICSs operated in isolated environments, but as they have become more connected to corporate networks and the internet, they are now exposed to a broader range of cyber threats.

NIST's Cybersecurity Framework emphasizes the need for ICS to be "protected through a combination of network segmentation, monitoring, and access control."[3] Network segmentation involves isolating ICS from other corporate networks to limit the potential damage in the event of a breach. Additionally, implementing multi-factor authentication (MFA) and role-based access controls can further limit access to sensitive systems.

Security of the Energy Supply Chain

The energy supply chain is another critical area of vulnerability, as attackers can target third-party vendors and service providers to gain access to energy systems. Cyberattacks on the energy supply chain can disrupt the production, transportation, and distribution of energy, leading to shortages and price spikes. For example, the Colonial Pipeline ransomware attack in 2021 caused major fuel shortages across the eastern United States, as the company was forced to shut down its operations to contain the attack.[4]

Securing the energy supply chain requires close collaboration between energy companies and their suppliers to ensure that cybersecurity standards are maintained throughout the entire chain. According to PwC, "a robust supply chain security program involves conducting regular risk assessments of third-party vendors, implementing security audits, and requiring compliance with industry standards such as NERC-CIP."[5] The North American Electric Reliability Corporation's Critical Infrastructure Protection (NERC-CIP) standards are a set of requirements designed to secure the systems that support North America's bulk electric system, including ensuring that third-party vendors comply with cybersecurity requirements.

Regulatory Frameworks and Compliance

In recognition of the critical importance of safeguarding energy infrastructure, governments around the world have implemented regulatory frameworks to ensure that energy and utility companies adopt strong cybersecurity practices. In the United States, the NERC-CIP standards are a key component of the regulatory landscape, requiring energy companies to implement security controls for critical systems, conduct vulnerability assessments, and report cybersecurity incidents. Failure to comply with NERC-CIP can result in significant fines, making compliance a top priority for energy companies.

In the European Union, the Network and Information Security (NIS) Directive aims to improve the security of critical infrastructure by requiring member states to implement measures for risk management and incident reporting in the energy, transport, and healthcare sectors. According to ENISA, "the NIS Directive has been instrumental in strengthening the cybersecurity posture of critical infrastructure operators in the EU, ensuring that they are better prepared to respond to cyber threats."[6] These regulatory frameworks are essential for maintaining the resilience of critical infrastructure and preventing the potentially catastrophic consequences of a cyberattack.

The Role of Cyber Resilience in the Energy Sector

Cyber resilience refers to an organization's ability to prepare for, respond to, and recover from cyberattacks while maintaining the continuity of critical operations. For energy and utility companies, cyber resilience is essential for ensuring that the lights stay on, even in the face of a cyberattack. A comprehensive cyber resilience strategy includes incident response planning, disaster recovery, and business continuity planning.

One of the key components of cyber resilience is the ability to detect and respond to cyber incidents in real time. Security operations centers (SOCs) play a crucial role in monitoring energy systems for signs of cyberattacks and coordinating responses across the organization. SOCs use security information and event management (SIEM) systems to aggregate and analyze data from multiple sources, enabling security teams to detect anomalies and potential threats. As noted by SANS Institute, "real-time threat detection and incident response are critical to minimizing the impact of cyberattacks on energy infrastructure."[7]

Protection of Renewable Energy Sources

As the energy sector transitions towards renewable energy sources such as wind, solar, and hydropower, new cybersecurity challenges are emerging. Renewable energy systems are highly distributed, with numerous small-scale energy producers connected to the grid. This increased complexity makes it more difficult to secure the entire system, as attackers have more entry points to exploit. Additionally, renewable energy systems rely heavily on smart grid technology, which uses sensors and communication networks to optimize energy distribution and consumption. While smart grids improve efficiency, they also introduce new cybersecurity vulnerabilities.

Securing renewable energy sources requires the implementation of cybersecurity best practices, such as encryption, access control, and continuous monitoring. According to Deloitte, "As renewable energy becomes an increasingly important part of the energy mix, energy companies should adopt advanced cybersecurity measures to protect these systems from cyberattacks." [8] Additionally, AI and machine learning are being used to enhance the cybersecurity of renewable energy systems by detecting and responding to threats in real time.

Conclusion

The energy and utilities sector is critical to the functioning of modern society, making it a prime target for cyberattacks. Protecting this infrastructure requires a multi-faceted approach that includes securing industrial control systems, safeguarding the energy supply chain, complying with regulatory frameworks, and enhancing cyber resilience. As the sector continues to evolve with the adoption of renewable energy and smart grid technologies, the need for advanced cybersecurity measures will only increase. By investing in cybersecurity and collaborating with government agencies and third-party vendors, energy and utility companies can strengthen their defenses and ensure the continuity of essential services.

6.4 Retail and E-Commerce: Ensuring Secure Transactions and Customer Trust

The retail and e-commerce sector has undergone a significant transformation in recent years, driven by the rise of online shopping and digital payment systems. While these changes have brought convenience to consumers, they have also introduced new cybersecurity challenges. Retailers handle vast amounts of sensitive customer data, including credit card information, personal details, and purchase histories. Protecting this data is critical not only for preventing financial losses but also for maintaining customer trust. The rise of cyberattacks targeting e-commerce platforms has made secure transactions a top priority for retailers. Implementing robust cybersecurity measures is essential for safeguarding both transactions and the customer relationship.

The Increasing Threat of Cyberattacks in Retail and E-Commerce

Retailers, particularly those operating online, are prime targets of cybercriminals due to the high volume of financial transactions they process daily. Common types of cyberattacks in the retail and e-commerce sector include phishing, payment fraud, data breaches, and distributed denial-of-service (DDoS) attacks. According to Fortinet's 2021 Retail Threat Landscape Report, "the retail sector experienced a 264% increase in cyberattacks in 2020, as attackers sought to exploit vulnerabilities in e-commerce platforms."[1] These attacks not only threaten the security of customer data but also disrupt business operations, leading to lost revenue and reputational damage.

One of the most notable examples of a cyberattack on the retail sector was the 2013 Target breach, where hackers stole credit card information from over 40 million customers by infiltrating the retailer's point-of-sale (POS) systems. This breach not only resulted in financial losses but also severely damaged the company's reputation, leading to a loss of customer trust [2]. Incidents like this demonstrate the importance of securing every aspect of the retail ecosystem, from online platforms to physical stores.

Protection of Customer Data by Use of Encryption and Tokenization

Securing customer data is the cornerstone of cybersecurity in retail and e-commerce. Given the sensitive nature of the data retailers collect, it is essential to protect this data using encryption and tokenization. Encryption ensures that data is converted into unreadable code when it is transmitted over networks, preventing unauthorized access. End-to-end encryption is widely used to secure transactions from the moment a customer submits payment information to when it is processed by the retailer.

Tokenization is another crucial tool for protecting customer data in the retail sector. Unlike encryption, tokenization replaces sensitive information with a randomly generated string of characters, or a "token," which can only be decoded by authorized systems. According to the PCI Security Standards Council, "tokenization is an effective method for reducing the risk of payment fraud, as it ensures that sensitive customer data is not stored in its original form." [3] This reduces the value of the data to cybercriminals, as tokens are useless without access to the decryption key.

Retailers that process online transactions are also required to comply with the Payment Card Industry Data Security Standard (PCI DSS), which mandates the implementation of security measures to protect cardholder data. Compliance with PCI DSS is essential for ensuring that payment transactions are secure and that customer data is protected from breaches. Failure to comply with PCI DSS can result in significant fines and the suspension of the retailer's ability to process credit card payments.

The Role of Multi-Factor Authentication (MFA) and Fraud Detection

One of the most effective ways to secure online transactions and protect customer accounts is through multi-factor authentication (MFA). MFA requires users to verify their identity using two or more methods—such as a password, a fingerprint, or a one-time code sent to their mobile device—before completing a transaction. This added layer of security makes it more difficult for cybercriminals to gain unauthorized access to customer accounts, even if they have stolen login credentials.

According to Verizon's 2021 Data Breach Investigations Report, "accounts protected by multi-factor authentication are significantly less likely to be compromised, as attackers should bypass multiple security barriers to gain entry."[4] MFA is particularly important in the retail sector where customers often use the same login credentials across multiple platforms, making them vulnerable to credential stuffing attacks.

In addition to MFA, retailers use fraud detection systems to monitor transactions for suspicious activity. These systems use machine learning and behavioral analysis to detect anomalies in customer behavior, such as unusual purchase patterns or transactions from unfamiliar locations. For example, if a customer's account is accessed from a foreign country shortly after a transaction in their home country, the system may flag the transaction as potentially fraudulent and require additional verification. As noted by Deloitte, "Advanced fraud detection systems help retailers identify and prevent fraudulent transactions before they occur, reducing the risk of financial losses."[5]

Security of E-Commerce Platforms and Mobile Applications

As e-commerce continues to grow, so too does the complexity of securing online platforms and mobile applications. Cybercriminals frequently target e-commerce websites with attacks such as SQL injection, cross-site scripting (XSS), and malware to exploit vulnerabilities and steal customer data. SQL injection involves inserting malicious code into a website's database query that allows attackers to gain access to sensitive information. Similarly, XSS attacks inject malicious scripts into web pages, which can then be executed by unsuspecting users.

To mitigate these risks, retailers should conduct regular security audits and implement vulnerability scanning to identify and address weaknesses in their platforms. According to NIST, "routine vulnerability assessments and patch management are essential for maintaining the security of e-commerce systems and preventing attackers from exploiting known vulnerabilities." [6] Retailers should also implement web application firewalls (WAFs) to filter and monitor HTTP traffic between web applications and the internet that would block potentially malicious traffic.

Mobile commerce (m-commerce) is another area of growing concern for cybersecurity in the retail sector. As more consumers use mobile apps to shop, retailers should ensure that their mobile applications are secure from cyberattacks. This includes encrypting data transmitted over mobile networks, securing APIs (application programming interfaces), and requiring customers to use strong, unique passwords. Mobile app security testing is also crucial for identifying and fixing vulnerabilities before they can be exploited by attackers.

Building Customer Trust Through Transparency and Privacy Protection

In addition to securing transactions, maintaining customer trust is critical for success in the retail and e-commerce sector. Transparency about data collection and privacy practices plays a key role in building and maintaining this trust. Customers want to know how their data is being used, who has access to it, and how it is being protected. As noted by PwC, "Retailers that are transparent about their data practices and prioritize customer privacy are more likely to retain customer loyalty, even in the event of a data breach."[7]

Retailers should comply with data protection regulations, such as the General Data Protection Regulation (GDPR) and the California Consumer Privacy Act (CCPA), which give consumers greater control over their personal data. These regulations require retailers to inform customers about the data they collect, obtain explicit consent for data processing, and provide mechanisms for customers to access or delete their data. Non-compliance with these regulations can result in substantial fines and damage to the retailer's reputation.

Incident Response and Cyber Resilience in Retail

Despite the best efforts to secure transactions and customer data, cyberattacks are inevitable. Therefore, having a robust incident response plan is essential for minimizing the impact of a breach and ensuring a swift recovery. An incident response plan outlines the steps that the retailer will take in the event of a cyberattack, from identifying the breach to containing the damage and notifying affected customers.

Retailer's ability to continue operating in the face of a cyberattack, minimizing disruptions to transactions and protecting customer trust is cyber resilience. According to SANS Institute, "Retailers that invest in cyber resilience are better positioned to respond to cyber incidents quickly and effectively, reducing the long-term impact on their business."[8] This includes implementing disaster recovery plans, data backups, and business continuity strategies to ensure that operations can continue even in the event of a major cyberattack.

Conclusion

Securing transactions and maintaining customer trust are critical priorities for retailers and e-commerce platforms. By adopting robust security measures, retailers can protect sensitive customer data and reduce the risk of cyberattacks. Compliance with data protection regulations and transparency about data privacy practices are also essential for building and maintaining customer trust. As the retail sector continues to grow in complexity, retailers should remain vigilant and invest in advanced technologies to strengthen their cybersecurity defenses and ensure the security of customer transactions.

6.5 Educational Institutions: Balancing Openness with Data Protection

The education sector faces unique cybersecurity challenges as it strives to maintain a balance between openness and data protection. Educational institutions, particularly universities, promote open access to information, encouraging collaboration and knowledge sharing. However, they also handle vast amounts of sensitive data, including student records, financial information, research data, and intellectual property. This openness, coupled with the vast digital footprints of students, faculty, and staff, makes educational institutions attractive targets for cyberattacks. As digital learning platforms and online educational tools grow, the need to protect both academic integrity and personal data has never been greater.

The Rising Threat of Cyberattacks in Educational Institutions

Educational institutions have become prime targets for cyberattacks due to their valuable data and relatively weak security defenses compared to other industries. Phishing attacks, ransomware, and data breaches are some of the most common threats faced by schools and universities. According to Verizon's 2022 Data Breach Investigations Report, "the education sector experienced a significant increase in ransomware attacks in 2021, with attackers often targeting student records and financial information."[1] These attacks can disrupt academic operations, compromise sensitive data, and erode trust between students, staff, and the institution.

In 2020, a major ransomware attack on Newhall School District in California forced the school to shut down its systems and disrupted online learning for thousands of students.[2] This incident highlighted the vulnerability of educational institutions to cyberattacks and the critical need for robust cybersecurity defenses to protect both educational continuity and sensitive data.

Protection of Student Data and Privacy

One of the primary cybersecurity concerns in education is the protection of student data. Schools and universities store a vast array of personally identifiable information (PII), including student names, addresses, Social Security numbers, and academic records. If this data falls into the wrong hands, it can be used for identity theft or other malicious purposes. FERPA (Family Educational Rights and Privacy Act) is a federal law in the United States that protects the privacy of student education records. FERPA requires schools to obtain consent before disclosing student information and mandates that schools take appropriate steps to protect these records.

Encryption and access controls are two critical tools for protecting student data. By encrypting data at rest and in transit, schools can ensure that sensitive information remains secure, even if it is intercepted by unauthorized users. Role-based access controls (RBAC) limit who can access certain data based on their role within the institution. For example, only administrators may have access to financial information, while teachers can access student grades. According to Educause, "the implementation of encryption and access controls is essential for safeguarding student data and complying with privacy regulations such as FERPA."[3]

In addition to protecting student data, educational institutions should ensure that digital learning platforms are secure. With the rise of online learning, schools now rely on platforms such as Zoom, Google Classroom, and Microsoft Teams to facilitate virtual classes and collaboration. These platforms should be secured to prevent unauthorized access, data breaches, and disruptions to online learning.

Balancing Openness and Security in Higher Education

One of the key challenges in the higher education sector is balancing openness with security. Universities are places of academic freedom and knowledge sharing, with researchers and students encouraged to collaborate openly across disciplines and institutions. However, this culture of openness can sometimes conflict with the need to protect sensitive data, intellectual property, and research findings. According to Schatz et al., "Universities should find a way to balance their core values of openness and academic freedom with the growing need for cybersecurity measures."[4]

Research universities are particularly vulnerable to cyberattacks due to the valuable data they produce, including cutting-edge research, intellectual property, and medical data from research trials. Cybercriminals, including nation-state actors, may target universities to steal this data for financial gain or to advance their own research efforts. In 2020, Michigan State University fell victim to a ransomware attack that compromised sensitive student and research data [5]. The attack demonstrated the risks associated with the open research environment and the need for universities to protect their intellectual property from cyber threats.

To address these challenges, universities should adopt a risk-based approach to cybersecurity, ensuring that they protect sensitive data while allowing for the open exchange of information where appropriate. This approach involves identifying and classifying data based on its sensitivity and implementing appropriate security controls based on the level of risk. Data classification tools can help universities categorize their data and apply the necessary security measures to protect it.

The Role of Multi-Factor Authentication (MFA) in Educational Security

Multi-factor authentication (MFA) is another key tool for protecting educational institutions from cyberattacks. MFA requires users to provide two or more forms of verification before accessing sensitive systems or data, making it much more difficult for cybercriminals to gain unauthorized access. Educational institutions, where students and staff often use the same passwords across multiple platforms, are particularly vulnerable to credential stuffing attacks, where attackers use stolen login credentials to access accounts.

Implementing MFA can significantly reduce the risk of such attacks. According to Deloitte, "Institutions that implement multi-factor authentication see a dramatic decrease in the number of successful account takeovers, as attackers are unable to bypass the additional layers of security."[6] MFA is especially important for securing access to systems that store sensitive student records, financial data, and research findings.

Ensure Cybersecurity in Online Learning

The shift to online learning during the COVID-19 pandemic accelerated the adoption of digital platforms and online tools in education, but it also exposed new cybersecurity vulnerabilities. Many schools and universities had to quickly transition to remote learning, often without fully vetting the security of the platforms they adopted. This rapid shift created opportunities for cybercriminals to exploit weaknesses in these systems, leading to incidents such as Zoom-bombing (the insertion of material that is lewd, obscene, or offensive in nature), where unauthorized users hijack online classes to disrupt learning.

To ensure the security of online learning environments, schools and universities should implement security best practices for digital platforms. This includes requiring strong passwords, enabling encryption for video calls, and restricting access to online classes to authenticated users. Privacy policies should also be clearly communicated to students, faculty, and parents, outlining how data collected during online learning will be used and protected.

According to The National Cyber Security Centre (NCSC), "schools should ensure that digital learning platforms are secure and that students' privacy is protected during online lessons."[7] By taking proactive steps to secure online learning environments, educational institutions can create a safer and more productive virtual learning experience.

Incident Response and Cyber Resilience

Despite the best efforts to secure educational institutions, cyberattacks are inevitable. Therefore, having a well-developed incident response plan is critical for minimizing the impact of a cyberattack and ensuring that academic operations can continue.

Cyber resilience is the ability of an educational institution to maintain operations in the face of a cyberattack. This includes implementing disaster recovery plans, data backups, and business continuity strategies to ensure that teaching and learning can continue even in the event of a major cyber incident. As noted by SANS Institute, "cyber resilience is essential for educational institutions to quickly recover from cyber incidents and maintain academic integrity."[8]

Conclusion

The education sector should strike a delicate balance between maintaining openness and ensuring robust data protection. As schools and universities continue to embrace digital tools and online learning, they should implement strong cybersecurity measures to protect student data, research, and intellectual property. By adopting encryption, access controls, multi-factor authentication, and incident response plans, educational institutions can safeguard sensitive information while fostering an environment of collaboration and academic freedom. Educational institutions should remain vigilant and invest in cybersecurity to protect their students, faculty, and the integrity of their academic mission.

6.6 Manufacturing: Securing Smart Factories and Industrial Control Systems

The manufacturing industry is undergoing a digital transformation with the rise of smart factories, industrial control systems (ICS), and Internet of Things (IoT) technologies. While these advancements have greatly improved efficiency, productivity, and automation, they have also introduced new vulnerabilities that make the industry an attractive target of cyberattacks. As manufacturing processes become more connected, the risk of attacks on industrial control systems and other critical infrastructure increases. Securing these systems is essential for protecting sensitive data, maintaining operational continuity, and ensuring the safety of workers and equipment.

The Rise of Smart Factories and Connected Systems

Smart factories are characterized by the use of IoT devices, robotics, and automated systems that communicate in real time to optimize production processes. These technologies enable manufacturers to increase efficiency, reduce costs, and improve product quality. However, they also create additional points of entry for cybercriminals. As noted by McKinsey, "smart factories are particularly vulnerable to cyberattacks due to their high level of interconnectivity and reliance on digital infrastructure."[1].

One of the primary risks in smart factories is the attack on industrial control systems (ICS), which include Supervisory Control and Data Acquisition (SCADA) systems, Distributed Control Systems (DCS), and Programmable Logic Controllers (PLCs). These systems control and monitor physical processes such as assembly lines, machine operations, and quality control. A cyberattack on an ICS can lead to significant disruptions, halting production, damaging machinery, and compromising product quality.

Threats to Industrial Control Systems (ICS)

Industrial control systems have traditionally operated in isolated environments, but as smart factories integrate them with broader corporate networks, they become more vulnerable to external threats. Cyberattacks on ICS can take many forms, including malware, ransomware, phishing, and denial-of-service (DoS). According to Dragos, "ICS are increasingly targeted by nation-state actors and cybercriminals seeking to disrupt critical infrastructure and industrial operations."[2]

The Stuxnet attack in 2010 is one of the most well-known examples of a cyberattack on an ICS. This attack demonstrated the potential for cyberattacks to cause real-world harm by targeting critical industrial processes.

To mitigate the risk of attacks on ICS, manufacturers should implement strong security controls that include network segmentation, regular security audits, and intrusion detection systems (IDS). Network segmentation involves isolating critical systems from broader corporate networks, reducing the likelihood of an attacker moving laterally across the network after gaining access. ID Scan monitors network traffic for suspicious activity, allowing security teams to respond quickly to potential threats.

Security of Industrial IoT (IIoT) Devices

As part of the shift towards smart factories, manufacturers are increasingly adopting Industrial Internet of Things (IIoT) devices, which connect machinery, sensors, and equipment to the Internet for real-time monitoring and control. IIoT devices offer significant benefits, such as predictive maintenance, where machines are serviced based on actual usage data rather than a fixed schedule, reducing downtime and extending equipment life.

However, the proliferation of IIoT devices also introduces new security challenges. Many IIoT devices have limited built-in security features, making them vulnerable to cyberattacks. According to Gartner, "the rapid deployment of IIoT devices in manufacturing has outpaced the implementation of cybersecurity measures, leaving many systems exposed to potential attacks."[3] To secure IIoT devices, manufacturers should implement encryption, authentication mechanisms, and regular firmware updates to patch vulnerabilities.

Cybersecurity Frameworks and Regulatory Compliance

Manufacturing companies are subject to various cybersecurity frameworks and regulations that aim to protect critical infrastructure and ensure the security of ICS and smart factory environments. In the United States, the NIST Cybersecurity Framework provides guidelines for managing and reducing cybersecurity risks in critical infrastructure sectors, including manufacturing. The framework emphasizes the importance of risk assessment, incident response, and continuous monitoring to detect and mitigate cyber threats.

Similarly, the IEC 62443 standard, developed by the International Electrotechnical Commission (IEC), provides a comprehensive framework for securing ICS and other operational technology (OT) environments. According to IEC, "the IEC 62443 standard is essential for ensuring that industrial control systems are protected from cyber threats throughout the entire product lifecycle."[4] Manufacturers should comply with these standards to ensure the security and resilience of their critical systems.

Prevention of Ransomware and Malware Attacks

Ransomware attacks are a growing threat to the manufacturing industry, as cybercriminals target ICS to demand ransom payments in exchange for restoring access to critical systems. A ransomware attack can halt production, resulting in significant financial losses and damage to a manufacturer's reputation. In 2019, Norsk Hydro, a global aluminum producer, was hit by a ransomware attack that forced the company to switch to manual operations for several weeks, costing the company an estimated $40 million.[5]

To prevent ransomware attacks, manufacturers should implement backup and recovery systems that allow them to restore operations without paying a ransom. Regular data backups, stored in secure, offsite locations, ensure that critical data can be recovered in the event of a ransomware attack. Additionally, manufacturers should implement email filtering and phishing awareness training to prevent employees from falling victim to phishing emails, which are often used to deliver ransomware.

Incident Response and Cyber Resilience in Manufacturing

Manufacturers should be prepared to respond quickly and effectively to cyberattacks to minimize downtime and ensure business continuity. A well-developed incident response plan should be implemented. According to SANS Institute, "Incident response planning is critical for manufacturers to quickly recover from cyberattacks and minimize the impact on production and supply chains."[6]

Cyber resilience should be embraced to ensure that production can continue even in the event of a major cyber incident. Manufacturers should regularly test their incident response plans through tabletop exercises and full-scale simulations to ensure that they are prepared to respond to real-world attacks.

Security of the Manufacturing Supply Chain

The manufacturing supply chain is another area of vulnerability, as attackers can target third-party suppliers to gain access to critical systems. Supply chain attacks can disrupt production, delay shipments, and compromise the quality of manufactured goods. For example, in 2020, a supply chain attack on SolarWinds, a software provider used by many manufacturers, allowed attackers to insert malware into a software update, compromising the systems of numerous organizations.[8]

To secure the manufacturing supply chain, manufacturers should conduct third-party risk assessments to evaluate the security practices of their suppliers. This includes ensuring that suppliers comply with relevant cybersecurity standards and implementing contractual agreements that require them to report cyber incidents. Additionally, manufacturers should implement supply chain monitoring tools to detect and respond to potential threats in real time.

Conclusion

The manufacturing industry is at the forefront of digital transformation, with smart factories and industrial control systems revolutionizing production processes. However, this increased connectivity also introduces new cybersecurity risks. By implementing robust security measures, manufacturers can protect their critical systems from cyberattacks. Additionally, compliance with cybersecurity frameworks like the NIST Cybersecurity Framework and IEC 62443 is essential for ensuring the security of industrial control systems and maintaining operational continuity. Manufacturers should remain vigilant and invest in advanced technologies to secure their smart factories and industrial environments.

Test Your Knowledge on Cybersecurity in Key Industries

Q1: What cybersecurity challenges are unique to the financial services?

Q2: How do healthcare organizations protect patient information from cyber threats?

Q3: What are the primary cybersecurity risks facing the energy and utilities sector?

Q4: How does cybersecurity support the protection of critical infrastructure in the energy sector?

Q5: What are the security concerns in the retail and e-commerce?

Q6: How do organizations in the education sector balance openness with data protection?

Q7: What are the main cybersecurity challenges in the manufacturing industry?

Q8: How does the healthcare industry address the cybersecurity risks posed by medical devices?

Q9: How do cybersecurity regulations impact the financial services sector?

Q10: Why is cybersecurity critical in safeguarding critical infrastructure sectors?

Q11: What role does encryption play in securing e-commerce transactions?

Q12: How does the education sector protect against cyberattacks while maintaining accessibility?

Q13: What are the cybersecurity risks associated with connected devices in the energy sector?

Q14: How does the retail industry ensure customer trust through cybersecurity?

Q15: What cybersecurity measures do smart factories require to prevent attacks?

ANSWERS

Q1: A: Challenges include protecting sensitive financial data, defending against fraud DDoS attacks, and maintaining compliance with strict regulations like PCI-DSS.

Q2: A: Healthcare organizations use encryption, access control, and data protection strategies to secure electronic health records (EHRs) and comply with HIPAA regulations.

Q3: A: The energy sector faces risks such as industrial control system (ICS) attacks, disruption of power grids, and targeted attacks from nation-state actors.

Q4: A: Cybersecurity ensures the availability, integrity, and confidentiality of systems that control power distribution, preventing disruptions caused by cyberattacks.

Q5: A: Retail and e-commerce companies should protect customer data, ensure secure transactions, and defend against payment card fraud and data breaches.

Q6: A: Educational institutions implement cybersecurity policies that allow open access to information while protecting sensitive student and staff data from breaches.

Q7: A: Challenges include securing industrial control systems, preventing intellectual property theft, and protecting smart factories from cyberattacks.

Q8: A: Healthcare organizations implement device encryption, software updates, and network monitoring to secure connected medical devices from cyber threats.

Q9: A: Financial services organizations should comply with regulations like PCI-DSS and GDPR, ensuring strong data protection measures are in place to avoid penalties.

Q10: A: Cybersecurity protects essential services, such as power grids and transportation systems, from disruptions that could have far-reaching consequences for society.

Q11: A: Encryption protects sensitive payment data during online transactions, preventing interception by malicious actors and ensuring customer trust.

Q12: A: The education sector uses role-based access control, firewalls, and data encryption to protect sensitive information while maintaining accessibility for students and staff.

Q13: A: Connected devices, such as smart meters and remote sensors, can be vulnerable to attacks if not properly secured, potentially leading to service disruptions.

Q14: A: Retailers implement strong encryption, secure payment gateways, and privacy policies to protect customer data, ensuring trust in their online transactions.

Q15: A: Smart factories need to implement network segmentation, secure communications, and monitoring systems to protect against cyber-physical attacks.

CHAPTER 7

THE ECONOMICS OF CYBERSECURITY

Note: All references in this chapter are on page 402 - 403

The financial consequences for companies and economies of cybersecurity, or lack thereof, will be carefully discussed in this chapter. First, we will examine the costs of cyberattacks—direct ones covering legal bills and ransom payments as well as indirect ones encompassing damage to reputation, loss of revenue, and long-term rehabilitation projects. Calculating the cost of security solutions against the probable savings from preventing invasions can help you understand how businesses determine the return on investment (ROI) for cybersecurity initiatives. Another crucial question is the purpose of cyber insurance since more businesses look for ways to spread the financial consequences of cyber occurrences. We will also go over the economics of cybercrime itself, including the dark web markets for traded viruses, stolen data, and hacking tools. This chapter looks at how companies may use smart cybersecurity investment plans and maximize cybersecurity expenditure to offer the best protection without depleting resources. You will learn how cybersecurity affects not just national economies but also corporate decisions and stock prices.

7.1 The Cost of Cyber Attacks on Businesses and Economies

Cyberattacks represent one of the most significant threats to the modern digital economy, affecting businesses of all sizes and industries. The financial impact of these attacks can be catastrophic, resulting in direct costs such as data loss, system downtime, legal fees, ransom payments, and indirect costs, including reputational damage and loss of customer trust. On a broader scale, cyberattacks also have a detrimental effect on national economies, disrupting key industries, damaging infrastructure, and undermining public confidence. Understanding the economic cost of cyberattacks is crucial for businesses and governments to implement effective cybersecurity strategies and mitigate the growing financial risks associated with cybercrime.

The Rising Financial Impact of Cybercrime

The cost of cyberattacks has grown exponentially in recent years as cybercriminals develop more sophisticated methods to breach security systems. According to Accenture's 2022 Cost of Cybercrime Study, "the global cost of cybercrime is expected to reach $10.5 trillion annually by 2025, driven by the increasing frequency and severity of attacks on businesses and governments."[1] This figure includes both direct financial losses and the broader economic damage caused by cyberattacks, such as reduced productivity, loss of intellectual property, and disruption to supply chains.

Businesses bear the brunt of these financial losses, with large corporations often paying millions of dollars to recover from data breaches, ransomware attacks, or denial-of-service (DoS) incidents. IBM's 2021 Cost of a Data Breach Report found that "the average cost of a data breach globally is $4.24 million, with healthcare, financial services, and energy sectors experiencing the highest financial impacts."[2] These costs include expenses related to detection and escalation, notification of affected individuals and entities, and legal and regulatory fines.

In addition to these direct costs, businesses also suffer from lost revenue and business disruption following a cyberattack. In the case of a ransomware attack, for example, companies may be forced to shut down operations for days or weeks while they attempt to recover systems or pay a ransom to regain access to critical data. This downtime can result in millions of dollars in lost sales and missed business opportunities, further compounding the financial damage.

Ransomware Attacks and Their Economic Impact

Ransomware attacks have emerged as one of the most financially devastating types of cyberattacks, with businesses across industries facing increasingly frequent and costly incidents. In a ransomware attack, cybercriminals encrypt an organization's data and demand a ransom in exchange for restoring access. If the ransom is not paid, the attackers may threaten to delete the data or release it publicly. According to Sophos' 2021 Ransomware Report, "the average cost to recover from a ransomware attack—including downtime, lost business, and ransom payments—was $1.85 million in 2021, a figure that has doubled in just a few years."[3]

Ransomware attacks not only result in direct costs, such as ransom payments and recovery efforts, but also cause reputational damage. When companies are forced to pay ransom, they often face criticism from customers, regulators, and the media for failing to adequately protect their data. This loss of trust can lead to a decline in customer retention and revenue over time. For example, after a high-profile ransomware attack on Colonial Pipeline in 2021, the company paid $4.4 million in ransom to restore operations, but the incident also led to widespread fuel shortages across the eastern United States, costing the economy billions of dollars in lost productivity. [4]

The Economic Ripple Effect of Cyberattacks on National Economies

The financial damage of cyberattacks extends beyond individual businesses to national and global economies. Major cyber incidents have the potential to disrupt critical infrastructure, such as energy grids, transportation systems, and healthcare services, causing cascading economic impacts. ENISA(European Union Agency for Cybersecurity) notes that "cyberattacks on critical infrastructure can have a multiplier effect on the economy, leading to widespread financial losses and reduced economic growth." [5]

In addition to infrastructure disruptions, the theft of intellectual property (IP) and trade secrets can have long-lasting economic consequences. When cybercriminals steal proprietary information or industrial designs, they can undermine a country's economic competitiveness and deprive businesses of the competitive advantage they have invested heavily in developing. McAfee's 2020 Economic Impact of Cybercrime Report states, "the global cost of IP theft due to cybercrime is estimated at over $500 billion annually, with sectors such as manufacturing, technology, and pharmaceuticals particularly affected."[6]

The Cost of Compliance and Regulatory Fines

In response to the growing threat of cyberattacks, governments around the world have implemented data protection regulations that impose significant fines on businesses that fail to adequately protect customer data. The General Data Protection Regulation (GDPR) in the European Union and the California Consumer Privacy Act (CCPA) in the United States are two prominent examples of laws that require businesses to adopt strict security measures and report data breaches in a timely manner.

Non-compliance with these regulations can result in hefty fines, further increasing the financial cost of a cyberattack. For example, under GDPR, companies can be fined up to 4% of their annual global revenue for failing to protect personal data or for failing to notify regulators of a data breach. In 2021, British Airways was fined £20 million for a data breach that compromised the personal information of over 400,000 customers[7]. Similarly, Equifax, a credit reporting agency, was fined $700 million following its 2017 data breach, which affected over 147 million people.[8]

In addition to fines, businesses should also invest heavily in compliance efforts to avoid regulatory penalties. This includes implementing security measures, conducting regular audits, and ensuring that employees are trained to recognize and respond to cyber threats. According to Deloitte, "the cost of complying with data protection regulations is rising, with businesses in highly regulated sectors such as finance and healthcare spending millions of dollars annually on compliance-related activities."[9]

Insurance and the Growing Demand for Cyber Coverage

Given the financial risks posed by cyberattacks, many businesses are turning to cyber insurance as a way to mitigate the potential costs of an attack. Cyber insurance policies typically cover expenses such as legal fees, data recovery, business interruption, and ransomware payments. However, as the frequency and severity of cyberattacks have increased, the cost of cyber insurance premiums has also risen sharply. According to Marsh McLennan, "cyber insurance premiums increased by 96% in 2021 due to the surge in ransomware attacks and the growing financial impact of data breaches."[10]

While cyber insurance can help businesses recover from the financial fallout of an attack, it is not a substitute for strong cybersecurity practices. Insurance providers are increasingly requiring companies to implement certain security measures—such as multi-factor authentication (MFA) and regular security audits—in order to qualify for coverage. Failure to meet these requirements can result in denied claims, leaving businesses to bear the full cost of an attack.

The Long-Term Economic Impact of Cyberattacks

The long-term economic impact of cyberattacks can be profound, affecting not only businesses but also entire industries and economies. In addition to the direct financial costs of recovering from an attack, businesses may face increased scrutiny from regulators, diminished customer trust, and higher insurance premiums. In some cases, companies may even be forced to shut down if the financial damage is too severe to recover from.

On a broader scale, cyberattacks can undermine investor confidence in critical industries, such as finance, energy, and healthcare, leading to reduced investment and slower economic growth. As PwC notes, "the financial risks associated with cyberattacks are now seen as a major threat to global economic stability, with the potential to disrupt markets and hinder economic development."[11] This highlights the importance of investing in cybersecurity at both the organizational and governmental levels to protect the global economy from the growing threat of cybercrime.

Conclusion

Cyberattacks have a significant economic impact on businesses and national economies, with costs that range from direct financial losses to long-term damage to reputation and customer trust. The rising frequency and sophistication of cyberattacks, particularly ransomware incidents, have made it more expensive than ever for businesses to protect themselves from cybercrime. As the global economy becomes increasingly interconnected, the financial risks posed by cyberattacks will continue to grow, underscoring the need for robust cybersecurity measures, regulatory compliance, and effective incident response strategies to mitigate the economic fallout of cybercrime.

7.2 Cybersecurity Investment Strategies and ROI

Investing in cybersecurity has become a top priority for businesses worldwide as the frequency and severity of cyberattacks continue to rise. Organizations should adopt cybersecurity investment strategies that not only protect their assets but also provide a clear return on investment (ROI). The cost of a cyberattack, including data breaches, ransomware, and downtime, can far outweigh the upfront investment in cybersecurity measures. Therefore, it is essential for businesses to implement effective cybersecurity strategies that align with their risk profile, budget, and overall business goals. Calculating the ROI of cybersecurity investments helps companies justify their spending on security tools and technologies, ensuring they allocate resources efficiently.

The Importance of Strategic Cybersecurity Investment

Strategic investment in cybersecurity is critical for mitigating cyberattack financial risks. According to Gartner, "the global spending on cybersecurity is expected to reach $188.3 billion by 2024, driven by the need to protect critical infrastructure, sensitive data, and intellectual property."[1] This spending includes investments in technologies such as firewalls, intrusion detection systems (IDS), encryption, and multi-factor authentication (MFA). In addition, it encompasses employee training, incident response plans, and cyber insurance, which are equally important for maintaining a strong security posture.

Many businesses face the challenge of determining how much to invest in cybersecurity and where to allocate those resources. This requires a risk-based approach that takes into account the organization's specific vulnerabilities, the potential impact of a cyberattack, and the cost-effectiveness of different security solutions. For example, a financial services company that handles sensitive customer data may prioritize investments in data encryption and network monitoring, while a manufacturing company might focus on securing its industrial control systems (ICS)to prevent operational disruptions.

Balancing Costs and Benefits: Measuring the ROI of Cybersecurity

Measuring the return on investment (ROI) for cybersecurity can be challenging, as it is often difficult to quantify the cost of an attack that has been prevented. However, businesses can calculate ROI by comparing the cost of implementing security measures with the potential financial losses they would incur from a cyberattack. According to IBM's 2021 Cost of a Data Breach Report, "the average cost of a data breach globally is $4.24 million, which includes expenses related to legal fees, regulatory fines, lost business, and recovery efforts."[2] Companies can avoid these significant financial losses by investing in cybersecurity measures that reduce the likelihood of a breach.

One way to measure the ROI of cybersecurity is to calculate the cost of risk reduction. This involves estimating the probability of a cyberattack and the potential financial impact of that attack, then determining how much can be saved by implementing specific security measures. For example, a company might determine that a data breach would cost $10 million and that investing $1 million in intrusion detection systems (IDS) can reduce the likelihood of a breach by 30%. In this case, the ROI of the investment would be the savings from avoiding the breach ($10 million x 30% = $3 million) minus the cost of the investment, resulting in a net ROI of $2 million. ei. ($10m x30% = $3m. $3m - $1m = $2m.

Cybersecurity as a Competitive Advantage

Beyond protecting assets and reducing financial risk, cybersecurity investments can also provide businesses with a competitive advantage. Customers, investors, and business partners increasingly prioritize companies that demonstrate a strong commitment to cybersecurity. According to a survey by PwC, "87% of consumers say they would take their business elsewhere if they don't trust a company's handling of their data."[3] By investing in robust cybersecurity measures and communicating those efforts to stakeholders, businesses can differentiate themselves in the marketplace and build customer trust.

In finance, healthcare, and e-commerce industries, where companies handle large volumes of sensitive data, demonstrating a strong security posture can be a key factor in winning contracts and retaining customers. For example, banks that invest in advanced fraud detection systems and end-to-end encryption may attract more customers by offering higher security for online transactions. Similarly, healthcare organizations that invest in HIPAA-compliant cybersecurity solutions can reassure patients that their medical records are safe from breaches.

The Role of Cyber Insurance in Cybersecurity Investment Strategies

As cyber threats adapt, many businesses are turning to cyber insurance to mitigate the financial impact of a successful attack. Cyber insurance policies can cover a range of costs, including ransomware payments, legal fees, business interruption, and data recovery. However, the cost of cyber insurance premiums has risen significantly in recent years due to the increasing frequency and severity of cyberattacks. According to Marsh McLennan, "cyber insurance premiums increased by 96% in 2021 as insurers faced higher payouts for ransomware attacks and data breaches."[4]

While cyber insurance can be a valuable part of a company's cyber risk management strategy, it is not a substitute for robust security measures. In fact, many insurers now require companies to implement certain security controls—such as multi-factor authentication (MFA) and regular security audits—in order to qualify for coverage. This has led businesses to invest in cybersecurity not only to protect their assets but also to reduce their insurance premiums. As Forrester Research notes, "Companies that implement strong cybersecurity measures can lower their cyber insurance costs by demonstrating a reduced risk profile."[5]

Investment in Employee Training and Awareness Programs

One of the most cost-effective cybersecurity investments is in employee training and awareness programs. Human error remains a leading cause of cyber incidents, with phishing attacks, weak passwords, and insecure practices accounting for a significant percentage of breaches. By training employees to recognize and respond to cyber threats, businesses can significantly reduce their risk of falling victim to attacks. The ROI of such programs can be substantial, as they help prevent breaches that could result in millions of dollars in losses.

Aligning Cybersecurity Investment with Business Objectives

To maximize the ROI of cybersecurity investments, businesses should align their security strategies with their overall business objectives. This requires a risk-based approach that focuses on the most critical assets and threats, ensuring that resources are allocated where they will have the greatest impact. For example, a retail company that processes thousands of online transactions daily may prioritize investments in payment security and fraud detection systems, while a technology company that relies on intellectual property may focus on protecting research and development data from theft.

As noted by McKinsey, "cybersecurity investment strategies should be tailored to the specific needs and risks of the business, with a focus on protecting the most valuable assets and ensuring business continuity."[7] By aligning cybersecurity investments with business goals, companies can ensure that they are not only reducing risk but also supporting growth and innovation.

Conclusion

Investing in cybersecurity is no longer an option for businesses—it is a necessity. As the cost of cyberattacks continues to rise, companies should adopt strategic cybersecurity investment strategies that protect their assets and deliver a clear return on investment. By calculating the ROI of cybersecurity investments, businesses can ensure that they are allocating resources efficiently and making decisions that align with their risk profile and business objectives. Additionally, investments in employee training, cyber insurance, and advanced security technologies can provide businesses with a competitive advantage, helping them build customer trust and reduce the financial impact of a cyberattack. Businesses and governments should invest in the right tools, cybersecurity innovations and technologies to stay ahead of attackers.

7.3 The Cyber Insurance Market and Risk Transfer

The growing frequency and complexity of cyberattacks have given rise to the cyber insurance market, providing businesses with a way to transfer cyber risk. Businesses increasingly recognize the need to protect themselves from the financial impact of breaches, ransomware attacks, and other security incidents. Cyber insurance serves as a vital tool in this risk management strategy and help businesses recover financially in the aftermath of a cyberattack. By offering coverage for a variety of expenses, including legal fees, data recovery, and business interruption costs, cyber insurance helps mitigate the financial damage associated with cyber incidents. However, the growing demand for cyber insurance has also led to higher premiums and stricter coverage requirements, creating challenges for both insurers and insured companies.

The Growth of the Cyber Insurance Market

The cyber insurance market has seen rapid growth in recent years, driven by the increasing number of high-profile cyberattacks and the corresponding need for businesses to manage their risk exposure. According to Allied Market Research, "the global cyber insurance market is projected to reach $28.6 billion by 2026, reflecting the rising demand for coverage across various industries."[1] This growth is particularly evident in sectors such as finance, healthcare, and energy, where they handle vast amounts of sensitive data and are frequently targeted by cybercriminals.

As the market for cyber insurance expands, insurers are offering more specialized coverage options tailored to the unique risks faced by different industries. For example, healthcare organizations may seek coverage for breaches involving patient data under HIPAA, while financial institutions may prioritize protection against payment fraud and regulatory fines. These tailored policies allow companies to address their specific vulnerabilities while ensuring that they are adequately covered in the event of a cyber incident.

Risk Transfer and the Role of Cyber Insurance

Risk transfer is a key component of cybersecurity strategies, allowing companies to shift some of the financial burden of a cyberattack to an insurer. By purchasing cyber insurance, businesses can transfer the risk of certain costs associated with a cyber incident, such as ransomware payments, legal fees, regulatory fines, and business interruption expenses. This helps to protect the company's financial stability and ensure that it can continue operations even in the face of a significant breach.

However, while cyber insurance can mitigate financial losses, it is not a substitute for robust cybersecurity practices. Insurers often require companies to demonstrate that they have implemented adequate security measures as a condition for coverage. Companies that fail to meet these requirements may face higher premiums or denied claims.

The Challenges of Underwriting Cyber Risk

One of the major challenges in the cyber insurance market is the difficulty of underwriting cyber risk. Unlike traditional forms of insurance, such as property or life insurance, where risks can be more easily quantified, cyber risk is constantly evolving. The dynamic nature of cyber threats, coupled with the interconnectedness of modern digital systems, makes it difficult for insurers to accurately assess the likelihood and potential impact of a cyberattack. As a result, insurers should rely on a combination of historical data, cyber risk assessments, and predictive modeling to determine premiums and coverage limits.

However, the rapid pace of technological change means that historical data may not always provide a reliable indicator of future risk. New vulnerabilities are constantly being discovered, and cybercriminals are continually developing more sophisticated attack methods. This uncertainty has led to higher premiums and more restrictive policy terms as insurers seek to protect themselves from potentially catastrophic losses. As PwC notes, "the challenge of underwriting cyber risk lies in the difficulty of predicting the next major attack and its potential financial impact."[2]

Cyber Insurance Exclusions and Coverage Limitations

While cyber insurance can provide valuable protection for businesses, it is important to understand the limitations and exclusions that are often included in policies. Many cyber insurance policies exclude coverage for certain types of incidents, such as insider threats, acts of terrorism, or nation-state attacks. Additionally, some policies may limit coverage for data restoration or business interruption, leaving companies to bear some of the costs themselves.

For example, the NotPetya attack in 2017, which was attributed to a nation-state actor, caused widespread damage to businesses around the world. However, many companies affected by the attack found that their cyber insurance policies excluded coverage for incidents classified as acts of war or terrorism, leaving them to shoulder the financial burden of recovery.[3] This has led to increased scrutiny of cyber insurance policies, with businesses now paying closer attention to the fine print to ensure that they are fully covered for the risks they face.

The Future of the Cyber Insurance Market

As the cyber insurance market continues to evolve, both insurers and businesses will need to adapt to the changing landscape of cyber risk. Insurers are increasingly using artificial intelligence (AI) and machine learning to improve their ability to assess cyber risk and develop more accurate pricing models. These technologies allow insurers to analyze vast amounts of data in real time, helping them identify emerging threats and better understand the risk profiles of their clients.

In addition to technological advancements, the market may also see increased government regulation and standardization of cyber insurance policies. In response to the growing financial impact of cyberattacks, some governments are considering the introduction of mandatory cyber insurance requirements for certain industries, particularly those that are considered critical to national security. According to Deloitte, "the future of the cyber insurance market will likely involve a greater role for government regulation, as policymakers seek to ensure that businesses are adequately protected from cyber risks."[4]

Conclusion

Cyber insurance has become an essential tool for businesses seeking to manage the financial risks associated with cyberattacks. By transferring some of the costs of a cyber incident to an insurer, companies can protect their financial stability and ensure that they can recover quickly from an attack. However, the rising cost of premiums and the challenges of underwriting cyber risk highlight the importance of maintaining strong cybersecurity practices. Businesses should carefully evaluate their insurance needs and ensure that their policies provide adequate coverage for the specific risks they face. As the cyber insurance market continues to grow and evolve, both insurers and businesses will need to remain agile in responding to the dynamic threat landscape.

7.4 Economic Incentives for Improving Cybersecurity

The need for robust cybersecurity measures becomes paramount, as the digital world grows increasingly complex. However, improving cybersecurity often requires significant financial investment. In this context, economic incentives have emerged as a critical tool for motivating businesses and governments to enhance their cybersecurity practices. These incentives are tax credits, subsidies, reduced insurance premiums and government grants. They are designed to encourage organizations to allocate resources to protect their networks, data, and infrastructure. By aligning financial interests with security goals, economic incentives can drive widespread adoption of cybersecurity best practices and reduce the overall risk of cyberattacks.

The Role of Financial Incentives in Cybersecurity Adoption

Businesses, particularly small and medium-sized enterprises (SMEs), often face financial constraints when it comes to investing in cybersecurity. Without sufficient resources, many organizations may delay or neglect the implementation of essential security measures, leaving themselves vulnerable to cyber threats. To address this issue, governments and industry bodies have introduced a range of economic incentives to encourage companies to prioritize cybersecurity.

One common form of economic incentive is the provision of tax credits for cybersecurity investments. According to PwC, "tax incentives are a powerful tool for encouraging businesses to invest in cybersecurity by offsetting the costs of implementing advanced security technologies."[1] For example, businesses that invest in encryption, firewalls, or intrusion detection systems (IDS) may be eligible for tax credits that reduce their overall tax liability. These incentives make cybersecurity investments more affordable and signal to businesses that security is a priority.

Government Subsidies and Grants for Cybersecurity Improvements

In addition to tax credits, government subsidies and grants are another form of economic incentive aimed at improving cybersecurity. Many governments offer financial assistance to businesses, especially in critical industries such as finance, healthcare, and energy, to help them adopt cutting-edge cybersecurity technologies and best practices. These subsidies can cover a range of expenses, from purchasing security tools to conducting vulnerability assessments and penetration testing.

For example, the European Union's Horizon 2020 program provides funding for cybersecurity research and development to enhance the security of critical infrastructure and emerging technologies such as artificial intelligence (AI) and quantum computing.[2] Similarly, the U.S. Department of Homeland Security (DHS) offers grants to help critical infrastructure operators improve their cybersecurity resilience through initiatives such as the Cybersecurity and Infrastructure Security Agency (CISA)'s Cybersecurity Grant Program. These grants reduce the financial burden on organizations, making it easier for them to invest in the security measures necessary to protect their operations from cyber threats.

Insurance Premium Discounts for Strong Cybersecurity Practices

Cyber insurance has become an increasingly important part of businesses' risk management strategies, providing coverage for the financial losses associated with data breaches, ransomware attacks, and other cyber incidents. To further incentivize businesses to invest in cybersecurity, many insurance providers offer premium discounts to companies that demonstrate a strong commitment to security. According to Marsh McLennan, "Businesses that implement advanced security measures, such as multi-factor authentication (MFA) and regular security audits, can qualify for lower cyber insurance premiums, reducing their overall insurance costs."[3]

This approach creates a financial incentive for businesses to enhance their security posture, as they can reduce their long-term insurance costs by demonstrating that they pose a lower risk to insurers. Moreover, insurers often work with businesses to identify gaps in their cybersecurity strategies and recommend improvements that can further reduce their exposure to cyber threats. As a result, the combination of insurance coverage and premium incentives encourage businesses to adopt a proactive approach to cybersecurity.

Public-Private Partnerships and Collaborative Efforts

Public-private partnerships(PPPs) are another important mechanism for fostering cybersecurity improvements. These collaborations between governments, businesses, and academic institutions create opportunities for knowledge sharing, research funding, and technological development aimed at strengthening cybersecurity across sectors. PPPs often provide financial incentives for private companies to participate in joint cybersecurity initiatives, such as developing new technologies or sharing threat intelligence with government agencies.

For example, the National Cyber Security Centre (NCSC) in the United Kingdom works with private companies to enhance the country's cybersecurity defenses through programs such as the Industry 100 initiative, which brings together industry experts and government officials to collaborate on cybersecurity challenges.[4] These partnerships not only foster innovation but also provide financial support to companies that contribute to the development of new security solutions.

The Role of Regulatory Incentives in Cybersecurity

Regulatory frameworks can also serve as economic incentives for improving cybersecurity. In many cases, compliance with cybersecurity regulations, such GDPR in the European Union or the CCPA in the United States, requires businesses to implement robust security measures. While these regulations are primarily designed to protect personal data, they also create financial incentives for compliance. Failure to comply with these regulations can result in substantial fines, making it financially prudent for businesses to invest in cybersecurity to avoid penalties.

For example, under GDPR, companies that fail to protect personal data can be fined up to 4% of their annual global revenue, creating a significant financial incentive to adopt data protection measures such as encryption, access controls, and incident response planning.[5] Similarly, businesses that comply with regulatory standards often benefit from increased customer trust, which can lead to higher revenues and a competitive advantage in the marketplace.

Encouraging Cybersecurity Research and Innovation

Economic incentives also play a key role in promoting cybersecurity research and innovation. Governments and industry organizations often offer research grants and financial awards to encourage the development of new technologies and solutions that can improve cybersecurity. These incentives help drive the creation of advanced tools and techniques for detecting and mitigating cyber threats, ultimately benefiting businesses and society as a whole.

For instance, the Defense Advanced Research Projects Agency (DARPA) in the United States funds cutting-edge research on topics such as automated cybersecurity, post-quantum cryptography, and AI-driven threat detection. These research efforts aim to address the cybersecurity challenges of the future and provide the private sector with the tools they need to protect against emerging threats.[6] By offering financial incentives for innovation, governments can accelerate the development of new cybersecurity solutions that keep pace with the evolving threat landscape.

Conclusion

Economic incentives are a critical tool for encouraging businesses and governments to invest in cybersecurity. By offering tax credits, subsidies, insurance premium discounts, and research grants, these incentives help reduce the financial barriers to implementing robust security measures. At the same time, regulatory incentives create a strong financial case for compliance, ensuring that businesses prioritize data protection to avoid costly fines. Economic incentives will play an increasingly important role in driving the widespread adoption of cybersecurity best practices and ensure the long-term resilience of the digital economy.

7.5 The Cybersecurity Job Market and Skills Gap

The cybersecurity job market has seen significant growth in recent years, driven by the rising number of cyberattacks and the increasing demand for skilled professionals to protect digital assets. However, despite the growing need for cybersecurity expertise, there is a pronounced skills gap that has left many businesses struggling to find qualified personnel. This shortage of skilled cybersecurity workers poses a serious threat to organizations and national security, as it limits their ability to respond to evolving cyber threats effectively. Addressing the cybersecurity skills gap requires a concerted effort from governments, educational institutions, and the private sector to increase training opportunities and develop the next generation of cybersecurity professionals.

The Growing Demand for Cybersecurity Professionals

As digital transformation accelerates across industries, the demand for cybersecurity professionals has skyrocketed. Organizations of all sizes are seeking experts who can secure their networks, protect sensitive data, and prevent cyberattacks. According to Cybersecurity Ventures, "the global demand for cybersecurity professionals is expected to reach 3.5 million unfilled positions by 2025, highlighting the severe shortage of talent in the field."[1] This demand is particularly strong in sectors such as finance, healthcare, and energy, where the consequences of a cyberattack can be devastating.

The rapid increase in cyberattacks, including ransomware and phishing, has further heightened the need for skilled cybersecurity workers. Businesses are increasingly investing in intrusion detection systems (IDS), firewalls, and encryption technologies, but these tools are only effective when managed by qualified professionals. Without a robust cybersecurity workforce, organizations remain vulnerable to attacks, which can result in significant financial losses, legal liabilities, and reputational damage.

The Cybersecurity Skills Gap

The cybersecurity skills gap refers to the mismatch between the number of available cybersecurity professionals and the growing demand for their expertise. This gap has been exacerbated by the rapid evolution of cyber threats, which require increasingly specialized skills and knowledge. According to a report by (ISC)[2], "nearly 60% of organizations report a shortage of cybersecurity staff, making it difficult to adequately defend against cyberattacks."[2] The skills gap is particularly acute in areas such as cloud security, penetration testing, and incident response, where organizations are struggling to find qualified candidates.

One of the primary reasons for the cybersecurity skills gap is the lack of formal education and training programs in the field. Many universities and colleges have been slow to develop comprehensive cybersecurity curricula, leaving students without the necessary skills to enter the workforce. As a result, businesses are often forced to rely on on-the-job training or hire professionals with unrelated backgrounds who should learn cybersecurity skills on the fly. According to Fortinet, "the lack of formal cybersecurity education is a major barrier to closing the skills gap, as it limits the pipeline of qualified candidates."[3]

The Impact of the Skills Gap on Businesses

The cybersecurity skills gap has serious implications for businesses, as it limits their ability to defend against cyber threats and respond to incidents in a timely manner. Without enough skilled personnel, organizations may struggle to implement effective security measures, monitor their networks for suspicious activity, or recover from attacks. According to a study by Accenture, "businesses that lack sufficient cybersecurity staff are more likely to experience data breaches and suffer greater financial losses as a result."[4]

The skills gap also leads to increased stress and burnout among existing cybersecurity professionals, who are often required to take on additional responsibilities to compensate for the lack of staff. This can lead to higher turnover rates and further exacerbate the shortage of skilled workers. In some cases, businesses may be forced to outsource their cybersecurity operations to third-party providers, which can be costly and may not provide the same level of protection as an in-house team.

Bridging the Cybersecurity Skills Gap

Addressing the cybersecurity skills gap requires a multifaceted approach that involves education, training, and professional development. One of the most important steps is to expand access to cybersecurity education at both the undergraduate and graduate levels. Universities and colleges should develop cybersecurity programs that provide students with the skills and knowledge they need to succeed in the field. These programs should focus on key areas such as network security, cryptography, incident response, and ethical hacking.

In addition to formal education, businesses can help bridge the skills gap by offering internship and apprenticeship programs that provide hands-on experience in cybersecurity. According to Deloitte, "Internships and apprenticeships are critical for developing practical skills in cybersecurity, as they allow students to apply their knowledge in real-world settings."[5] These programs also give businesses the opportunity to identify and recruit talented individuals before they enter the job market.

Certifications and Continuous Learning

In the rapidly evolving field of cybersecurity, continuous learning is essential for staying up to date with the latest threats and technologies. Professional certifications, such as Certified Information Systems Security Professional (CISSP), Certified Ethical Hacker (CEH), and CompTIA Security+, etc. provide individuals with the specialized skills needed to succeed in the industry. According to Gartner, "certifications play a key role in closing the cybersecurity skills gap by providing a standardized measure of expertise that employers can rely on."[6]

Businesses can also encourage continuous learning by offering training programs and professional development opportunities for their existing cybersecurity staff. This not only helps employees stay current with the latest trends but also boosts job satisfaction and reduces turnover. Investing in the professional development of cybersecurity staff can help businesses retain their top talent and ensure that they are prepared to defend against emerging threats.

The Role of Government and Industry in Closing the Skills Gap

Governments and industry organizations also have a critical role to play in closing the cybersecurity skills gap. Many governments have launched initiatives aimed at increasing the number of trained cybersecurity professionals, particularly in critical sectors such as defense, energy, and healthcare. For example, the U.S. Department of Homeland Security (DHS) offers scholarships and grants to students pursuing degrees in cybersecurity, with the goal of building a pipeline of talent for the public and private sectors.[7]

Industry organizations, such as (ISC)², E-Council and CompTIA, have also developed programs to address the skills gap. These programs include workshops, training sessions, and certification exams designed to equip individuals with the skills they need to succeed in the field. By collaborating with educational institutions and businesses, these organizations are helping to build a more robust cybersecurity workforce.

Conclusion

The cybersecurity skills gap represents a significant challenge for businesses and governments worldwide as the demand for qualified professionals continues to outpace supply. To address this gap, it is important to invest in education, training, and professional development programs that provide individuals with the skills they need to succeed in the field. By expanding access to cybersecurity education, offering internships and apprenticeships, and promoting continuous learning, businesses and governments can help close the skills gap and build a more resilient cybersecurity workforce. Addressing the skills gap is critical to ensuring the long-term security of digital infrastructure and protecting businesses from the growing risks of cyberattacks.

7.6 Cybercrime Economics and the Dark Web Marketplace

The rise of cybercrime has created an underground economy where stolen data, malware, and illegal services are bought and sold. This marketplace, often referred to as the dark web, is a hidden part of the internet where anonymity is maintained through the use of encryption and virtual private networks (VPNs). The dark web serves as a hub for cybercriminals to trade in illicit goods and services, ranging from stolen credit card numbers to ransomware kits and hacking services. Understanding the economics of cybercrime is crucial for businesses and governments to combat these threats effectively and implement cybersecurity strategies that can mitigate their impact.

The Growth of the Dark Web Marketplace

The dark web has grown into a thriving marketplace, enabling cybercriminals to monetize their activities. According to The Rand Corporation, "the dark web has evolved into a global network where criminals can buy and sell tools, data, and services for cyberattacks."[1] The anonymity provided by the dark web allows cybercriminals to operate with little fear of detection or prosecution, making it an attractive platform for illicit trade.

Cybercriminals on the dark web offer a wide range of products and services, including Ransomware as a service (RaaS), phishing kits, and stolen identities. These items are sold for cryptocurrency, such as Bitcoin, which further enhances anonymity. The dark web also facilitates the exchange of zero-day vulnerabilities, which are software flaws unknown to developers that can be exploited by attackers before a patch is available. These vulnerabilities are highly sought after and can be sold for substantial sums, sometimes in the millions of dollars.

The Economics of Cybercrime

The dark web's marketplace operates much like a legitimate economy, with buyers, sellers, and intermediaries engaging in transactions based on supply and demand. Cybercrime has become a lucrative business, generating billions of dollars in revenue each year. According to McAfee's 2020 Economic Impact of Cybercrime Report, "the global cost of cybercrime is estimated at $1 trillion annually, driven by the increasing prevalence of ransomware, data breaches, and financial fraud."[2]

The economic model of cybercrime is based on the commoditization of data and malware. For example, stolen credit card numbers are often sold in bulk on the dark web for as little as $10 per card, while personal identification information (PII), such as Social Security numbers or medical records, can fetch higher prices due to their utility in committing identity theft and financial fraud. Ransomware kits, which allow even non-technical criminals to launch attacks, are sold on a subscription basis, enabling widespread use of this devastating form of malware.

One of the key drivers of the dark web economy is ransomware, which has become one of the most profitable forms of cybercrime. As Sophos' 2021 Ransomware Report notes, "the average ransom paid by businesses to recover their data increased by 171% in 2021, reaching $312,493."[3] Ransomware attacks often involve double extortion, where attackers not only encrypt the data but also threaten to release it publicly unless the ransom is paid.

The Role of Cryptocurrencies in the Dark Web Economy

Cryptocurrencies play a central role in the dark web economy by providing a relatively anonymous method for cybercriminals to conduct financial transactions. The use of Bitcoin and other cryptocurrencies allows cybercriminals to buy and sell goods and services without revealing their identities or locations. This has made it difficult for law enforcement agencies to trace transactions and apprehend criminals operating on the dark web.

According to Chainalysis' 2021 Crypto Crime Report, "the use of cryptocurrencies in cybercrime-related activities, including ransomware payments and dark web transactions, has grown significantly in recent years, accounting for over $10 billion in illicit transactions in 2020 alone."[4] While cryptocurrencies offer certain levels of anonymity, they are not entirely untraceable. Blockchain analysis tools have been developed to help trace cryptocurrency transactions, and law enforcement agencies are increasingly using these tools to track and disrupt criminal networks operating on the dark web.

Ransomware as a Service (RaaS): The Franchising of Cybercrime

One of the most significant developments in the cybercrime economy is the rise of Ransomware as a Service (RaaS). RaaS allows cybercriminals to "franchise" their ransomware operations, offering malware and support services to less-skilled criminals in exchange for a share of the profits. This model has lowered the barrier to entry for cybercrime, enabling a wider range of individuals to launch ransomware attacks without needing advanced technical knowledge.

As noted by Check Point, "Ransomware as a Service has transformed the ransomware landscape, making it easier for cybercriminals to launch attacks and increase the overall volume of attacks globally."[5] RaaS operators typically provide a user-friendly interface and customer support to help affiliates deploy ransomware campaigns, collect payments, and decrypt victims' data. In return, the affiliates pay a percentage of the ransom proceeds to the RaaS operators.

The profitability of the RaaS model has led to a surge in ransomware attacks across industries, from healthcare to education. This has created a vicious cycle where the success of ransomware attacks encourages more criminals to join the dark web marketplace, further fueling the growth of the cybercrime economy.

Efforts to Combat the Dark Web and Cybercrime

Governments and law enforcement agencies around the world are working to combat the dark web and disrupt the cybercrime economy. Efforts to shut down dark web marketplaces, such as the takedown of Silk Road in 2013 and AlphaBay in 2017, have dealt significant blows to the illegal trade of drugs, weapons, and cybercrime-related goods. However, new marketplaces continue to emerge, and cybercriminals are finding increasingly sophisticated ways to evade detection.

International cooperation is critical for combating cybercrime on the dark web, as criminals often operate across borders. Agencies such as Europol and the U.S. Federal Bureau of Investigation (FBI)have led efforts to coordinate global investigations and prosecute individuals involved in cybercrime. According to Interpol, "international collaboration is essential for tracking and dismantling dark web marketplaces, as these networks often span multiple countries and jurisdictions."[6]

In addition to law enforcement efforts, the private sector is also playing a role in disrupting the cybercrime economy. Cybersecurity firms are developing advanced tools for detecting and preventing cyberattacks, while Blockchain analysis companies are working to trace cryptocurrency transactions linked to criminal activity. By working together, governments, law enforcement, and private companies can make it more difficult for cybercriminals to profit from their activities.

Conclusion

The dark web marketplace has created a thriving underground economy where cybercriminals can buy and sell the tools needed to launch attacks, steal data, and extort victims. The commoditization of malware, stolen data, and hacking services has fueled the growth of the cybercrime economy, which now generates billions of dollars in revenue each year. While cryptocurrencies have made it easier for cybercriminals to conduct transactions anonymously, efforts to trace these transactions and disrupt dark web marketplaces are gaining momentum. Governments, law enforcement, and private sector actors should continue to collaborate in their efforts to combat cybercrime and dismantle the economic infrastructure that supports it.

Test Your Knowledge On The Economics of Cybersecurity

Q1: How do cyberattacks impact businesses and the global economy?

Q2: What are the direct and indirect costs associated with cyberattacks?

Q3: How do organizations calculate the return on investment (ROI) for cybersecurity?

Q4: What role does cyber insurance play in mitigating financial risks?

Q5: What factors influence the cost of cyber insurance premiums?

Q6: How can organizations make informed decisions about cybersecurity investments?

Q7: How do cyberattacks affect stock prices and shareholder value?

Q8: How do businesses justify increased spending on cybersecurity?

Q9: What is the role of government regulations in influencing cybersecurity investments?

Q10: How does the cost of cybersecurity breaches vary across industries?

Q11: How do ransomware attacks influence cybersecurity spending?

Q12: What strategies can businesses adopt to reduce cybersecurity costs without compromising security?

Q13: How do economic incentives promote stronger cybersecurity practices?

Q14: What is the role of the cybersecurity job market in economic growth?

Q15: How does the cybercrime economy operate on the dark web?

ANSWERS

Q1: A: Cyberattacks cause significant financial losses through data breaches, operational disruptions, legal fees, and reputational damage, impacting both individual businesses and the global economy.

Q2: A: Direct costs include ransom payments, legal fees, and incident response, while indirect costs involve downtime, loss of customer trust, and long-term reputational damage.

Q3: A: ROI is calculated by assessing the potential financial losses from a cyberattack compared to the costs of implementing preventive security measures.

Q4: A: Cyber insurance helps cover financial losses resulting from cyber incidents, such as data breaches and ransomware attacks, offering financial protection for affected organizations.

Q5: A: Factors include the organization's size, industry, past security incidents, existing cybersecurity measures, and the required coverage level.

Q6: A: Organizations can assess risks, evaluate potential threats, conduct cost-benefit analyses, and prioritize investments in technologies that offer the highest security ROI.

Q7: A: Cyberattacks often lead to immediate declines in stock prices due to loss of investor confidence, with long-term effects on shareholder value if trust isn't restored.

Q8: A: Businesses justify increased cybersecurity spending by demonstrating the potential savings from preventing breaches, avoiding regulatory fines, and protecting intellectual property.

Q9: A: Regulations such as GDPR and HIPAA mandate strong cybersecurity practices, encouraging businesses to invest in security to remain compliant and avoid fines.

Q10: A: Costs vary depending on the industry, with sectors like healthcare and financial services facing higher costs due to stringent regulatory requirements and the sensitive nature of their data.

Q11: A: The rise in ransomware attacks has prompted organizations to increase spending on preventive measures, such as endpoint protection and employee training, to avoid potential ransom payments.

Q12: A: Businesses can adopt cost-effective strategies like cloud-based security solutions, outsourcing to managed security service providers (MSSPs), and investing in automated security tools.

Q13: A: Governments and industry bodies may offer incentives, such as tax credits or grants, to organizations that invest in robust cybersecurity measures, encouraging widespread adoption.

Q14: A: The demand for cybersecurity professionals contributes to economic growth by creating high-paying jobs, fostering innovation in security technologies, and ensuring safer digital ecosystems.

Q15: A: The dark web serves as a marketplace for cybercriminals to buy and sell stolen data, hacking tools, and malicious services, fueling the global cybercrime economy.

CHAPTER 8

ETHICAL AND LEGAL ASPECTS OF CYBERSECURITY

Note: All references in this chapter are on page 403 - 404

Cybersecurity becomes more and more important for protecting systems and data; likewise, the need of negotiating the ethical and legal complexity related to it. This chapter explores the important ethical issues including the balance between security and privacy, the function of government monitoring, and business accountability following data breaches.

This chapter explores key ethical issues around security and privacy. It covers government monitoring and business accountability after data breaches. We will discuss ethical hacking and responsible disclosure practices. Cybersecurity experts find and report flaws before attackers do. The chapter explains the legal landscape, including the GDPR regulations. We will explore international cybersecurity laws and compliance standards. Companies should adapt to avoid risky areas of activity. AI-driven security systems raise concerns about privacy, discrimination, and transparency. The chapter emphasizes the ethical consequences of such technologies. By the end, you'll understand how ethics and law shape cybersecurity.

8.1 Privacy vs. Security: Striking the Right Balance

Now the clash between security and privacy is among the most highly debated issues of the digital age. As cybersecurity risks becoming more complex and pervasive, governments, businesses, and people fight to reconcile the desire for security with the protection of personal privacy. Striking the right balance is critical to ensuring both the safety of data and the protection of personal privacy. While security measures are essential for protecting against cyberattacks, they can also infringe on individual privacy if not implemented appropriately. Finding the equilibrium between these competing interests is a complex and ongoing challenge that requires careful consideration of ethical, legal, and technological factors.

The Increasing Threat of Cyberattacks and the Need for Security

The rise of cyberattacks has heightened the need for robust security measures to protect sensitive data, critical infrastructure, and national security. Cyberattacks such as ransomware, phishing, and data breaches have become more frequent and sophisticated, targeting both private businesses and government agencies. Accenture's 2022 Cybercrime Report states, "the number of cyberattacks increased by 31% in 2021, with businesses and governments facing mounting pressure to enhance their security defenses."[1] This surge in cyber threats has led to increased surveillance, data collection, and security protocols designed to detect and prevent attacks.

However, these heightened security measures often come at the expense of individual privacy. Governments and organizations frequently collect vast amounts of personal data, such as browsing histories, location data, and communication records, to monitor for suspicious activity. While such data collection can be an effective tool for identifying potential threats, it raises concerns about how personal information is accessed, stored, and shared without consent.

The Ethical Dilemma: Balancing Privacy and Security

The ethical dilemma surrounding privacy and security lies in the conflict between two fundamental rights: the right to personal privacy and the need for collective security. On the one hand, individuals have the right to control their personal information and decide how it is used. This right is enshrined in laws such as the General Data Protection Regulation (GDPR) in the European Union, which gives individuals control over their data and requires organizations to protect it.

On the other hand, governments and businesses are responsible for protecting their citizens and customers from cyber threats, which may require collecting and analyzing personal data. According to Solove and Schwartz, "security measures that encroach on privacy are often justified by the need to prevent harm, but they should be carefully balanced to avoid infringing on personal freedoms."[2] The challenge is determining how much privacy should be sacrificed in the name of security and where to draw the line between legitimate security measures and unjustified intrusions into private life.

The Role of Data Encryption in Balancing Privacy and Security

Data encryption is one of the key tools that can help balance privacy and security. Encryption ensures that data is rendered unreadable to unauthorized users, allowing individuals and organizations to protect sensitive information from cyberattacks. At the same time, encryption can limit the ability of governments and law enforcement agencies to access data in cases where it is needed for criminal investigations or national security.

The debate over encryption came to the forefront in 2016 when the FBI sought to compel Apple to unlock an iPhone used by one of the attackers in the San Bernardino shooting. Apple refused, arguing that creating a backdoor for the FBI would weaken encryption for all users and potentially compromise the privacy and security of millions of iPhone owners.[3] This case highlighted the tension between privacy and security, with both sides presenting compelling arguments. While the FBI argued that access to the data was essential for national security, privacy advocates warned that weakening encryption could set a dangerous precedent and lead to mass surveillance.

Government Surveillance and Privacy Concerns

Government surveillance is a major area where the conflict between privacy and security is most evident. In the aftermath of events such as the 9/11 attacks and the rise of global terrorism, many governments expanded their surveillance capabilities to monitor communications and detect potential threats. Programs like the USA PATRIOT Act in the United States gave law enforcement agencies broad powers to collect and analyze data from citizens without their knowledge or consent.

While such surveillance measures are often justified as necessary for national security, they raise significant privacy concerns. According to The Electronic Frontier Foundation (EFF), "government surveillance programs, particularly those conducted without oversight or transparency, risk infringing on individual privacy rights and creating a culture of mass surveillance."[4] The Snowden revelations in 2013 exposed the extent of government surveillance programs, revealing that agencies such as the NSA had been collecting vast amounts of data from individuals around the world, often without their consent.

In response to these concerns, many countries have introduced legislation aimed at limiting government surveillance and protecting privacy. For example, the European Court of Justice ruled that the EU-US Privacy Shield, which allowed for the transfer of personal data between the EU and the US, was invalid because it did not provide adequate protection against US government surveillance.[5] This decision was made in a case known as "Schrems II" in 2020 and the ruling underscored the importance of privacy protections in an increasingly interconnected world.

Striking the Right Balance: Legal and Ethical Frameworks

Striking the right balance between privacy and security requires the development of legal and ethical frameworks that address the competing interests at play. Laws such as GDPR and the California Consumer Privacy Act (CCPA) have been implemented to give individuals greater control over their personal data while also holding organizations accountable for protecting that data. It should be noted that these two laws were there before "Schrems II" in 2020. These laws require companies to obtain consent before collecting data, provide transparency about how data is used, and implement security measures to prevent data breaches.

However, these frameworks should also consider the legitimate needs of governments and businesses to protect against cyber threats. As noted by Cavoukian and Jonas, "privacy and security are not mutually exclusive; with the right policies in place, organizations can achieve both."[6] This can be achieved through the principle of privacy by design, which advocates for building privacy protections into the design of systems and technologies from the outset rather than treating privacy as an afterthought.

Conclusion

The tension between privacy and security is unlikely to disappear, as both are fundamental to the functioning of modern society. While security measures are essential for protecting individuals and organizations from cyber threats, they should be implemented in ways that respect privacy rights. By developing legal and ethical frameworks that balance these competing interests, governments and businesses can ensure that they are protecting both the safety and the freedoms of their citizens and customers. As the digital world continues to evolve, finding the right balance between privacy and security will remain a critical challenge for policymakers, technologists, and society as a whole.

8.2 Ethical Hacking and Responsible Disclosure

Ethical hacking, also known as white-hat hacking, plays a critical role in cybersecurity by identifying vulnerabilities in systems before malicious hackers can exploit them. Unlike their criminal counterparts, ethical hackers are authorized by organizations to perform penetration testing, network analysis, and system diagnostics to strengthen cybersecurity defenses. The practice of responsible disclosure, whereby vulnerabilities discovered by ethical hackers are reported to the affected organization or software vendor, is essential for improving security while minimizing harm. Ethical hacking and responsible disclosure help prevent cyberattacks, safeguard data, and ensure the protection of sensitive systems.

The Role of Ethical Hacking in Cybersecurity

Ethical hackers are hired by organizations to identify and exploit vulnerabilities in a controlled environment. This allows companies to fix security flaws before black-hat hackers discover them. According to Cybersecurity Ventures, "the demand for ethical hackers has grown significantly as organizations seek proactive approaches to protect their systems from cyber threats."[1] Ethical hackers use the same tools and techniques as malicious hackers but within legal and ethical boundaries, adhering to pre-agreed rules of engagement.

The importance of ethical hacking in today's cybersecurity landscape cannot be overstated. As McKinsey notes, "ethical hackers help companies build more resilient systems by simulating real-world attacks, giving organizations valuable insights into their security weaknesses"[2]. By identifying vulnerabilities such as SQL injections, cross-site scripting (XSS), or misconfigured firewalls, ethical hackers enable organizations to implement effective defenses and mitigate the risks associated with cyberattacks.

Responsible Disclosure: Ethics and Best Practices

The concept of responsible disclosure involves ethical hackers reporting vulnerabilities to the affected parties and allowing them time to fix the issues before making the findings public. This process ensures that security flaws are addressed in a timely manner while minimizing the risk that malicious actors will exploit the vulnerability before it is patched. According to Bruce Schneier, "responsible disclosure creates a win-win situation where organizations can secure their systems without exposing them to immediate risk, while ethical hackers receive recognition for their contributions to improving cybersecurity."[3]

Many organizations implement bug bounty programs, which offer financial rewards to ethical hackers who discover and report vulnerabilities. These programs provide incentives for hackers to act responsibly and report security issues through official channels rather than exploiting or selling the vulnerabilities. Companies such as Google, Microsoft, and Facebook have implemented successful bug bounty programs, paying out millions of dollars to ethical hackers who help improve their security.

However, responsible disclosure requires careful coordination and communication. The organization receiving the vulnerability report should acknowledge the issue, prioritize the patching process, and work with the ethical hacker to ensure that the flaw is addressed before public disclosure. CERT/CC (Computer Emergency Response Team Coordination Center) often facilitates responsible disclosure by coordinating communications between ethical hackers and affected vendors, particularly in cases where the vulnerability affects multiple organizations or software platforms.[4]

The Legal and Ethical Challenges of Ethical Hacking

While ethical hacking is a powerful tool for improving cybersecurity, it presents several legal and ethical challenges. One key challenge is ensuring that ethical hackers operate within legal boundaries. In many jurisdictions, hacking—regardless of intent—is illegal unless it is explicitly authorized by the organization being tested. Kevin Mitnick, a former hacker turned cybersecurity consultant, noted, "Even well-intentioned ethical hackers should obtain clear and explicit permission from the organizations they test to avoid legal repercussions"[5].

Another challenge is the potential for unintentional damage during penetration testing. Even with the best intentions, ethical hackers can inadvertently disrupt systems, cause data loss, or compromise operations, especially when testing critical infrastructure such as power grids or healthcare systems. Organizations should work closely with ethical hackers to define the scope of the testing, set clear boundaries, and ensure that appropriate safeguards are in place to minimize the risk of unintended harm.

The Role of Bug Bounty Programs in Ethical Hacking

Bug bounty programs have become an increasingly popular way for organizations to engage with ethical hackers and encourage responsible disclosure. These programs allow ethical hackers to submit vulnerability reports in exchange for financial rewards. According to HackerOne, a leading bug bounty platform, "bug bounty programs have paid out over $100 million to ethical hackers globally, demonstrating their effectiveness in improving cybersecurity through crowd-sourced vulnerability discovery."[6]

Bug bounty programs offer several advantages. First, they provide organizations with access to a large pool of skilled security researchers, often identifying vulnerabilities that internal teams might overlook. Second, they offer an incentive for hackers to act ethically, as they can earn money and recognition by reporting vulnerabilities rather than exploiting them. However, bug bounty programs also come with challenges, including managing the influx of vulnerability reports and determining appropriate rewards for each discovery.

The Impact of Ethical Hacking on Public Trust

Ethical hacking and responsible disclosure also play a crucial role in building public trust in digital systems. When organizations demonstrate a commitment to addressing vulnerabilities and engaging with the cybersecurity community, they reassure customers and stakeholders that their data is being protected. As Forrester Research noted, "Organizations that participate in bug bounty programs and practice responsible disclosure build a reputation for transparency and accountability, enhancing customer trust."[7]

This is particularly important in industries such as finance, healthcare, and government, where the consequences of a security breach can be devastating. Ethical hacking helps identify vulnerabilities before they can be exploited by malicious actors, thereby preventing breaches that could lead to financial loss, identity theft, or disruptions to critical services.

Ethical Hacking in Critical Infrastructure

One of the most important areas where ethical hacking is making an impact is in the protection of critical infrastructure. As systems such as power grids, transportation networks, and water treatment plants become more connected to the Internet, they are increasingly vulnerable to cyberattacks. Ethical hackers play a vital role in securing these systems by identifying vulnerabilities before nation-state actors or cyber criminals can exploit them.

In 2017, for example, ethical hackers discovered vulnerabilities in the US power grid that could have been exploited to cause widespread disruptions. By reporting these vulnerabilities through responsible disclosure, the hackers helped the government strengthen its defenses and prevent a potential attack.[8] This underscores the importance of ethical hacking in protecting national security and ensuring the continued operation of critical infrastructure.

Conclusion

The role of ethical hacking and responsible disclosure will become even more critical in safeguarding systems and data because of the evolving nature of cyber threats. Ethical hackers provide invaluable insights into vulnerabilities, allowing organizations to stay ahead of malicious actors. The cybersecurity community, governments, and businesses should continue to collaborate to establish clear legal and ethical frameworks that support ethical hacking while protecting against unintended harm. By fostering a culture of responsible disclosure, organizations can build more secure systems and earn the trust of their customers and stakeholders.

8.3 Government Surveillance and Civil Liberties

Government surveillance has become a central issue in the digital age, particularly in the context of national security and cybersecurity. While surveillance is often justified as a necessary tool for preventing terrorism, cyberattacks, and other threats, it raises significant concerns about civil liberties, including the right to privacy and freedom from unwarranted government intrusion. Striking the right balance between national security and protecting civil liberties has proven to be one of the most contentious debates in modern democratic societies. The expansion of surveillance programs in the wake of terrorist attacks and the growing capabilities of surveillance technologies have intensified this debate, with both supporters and critics weighing in on the ethical and legal implications of government surveillance.

The Role of Government Surveillance in National Security

Governments worldwide have expanded their surveillance capabilities in response to growing threats from cybercrime, terrorism, and espionage. Surveillance is often seen as a necessary tool for monitoring suspicious activities, detecting potential attacks, and gathering intelligence on hostile actors. Programs such as the USA PATRIOT Act, enacted in the wake of the 9/11 attacks, granted US law enforcement agencies broad powers to collect data on individuals suspected of being involved in terrorist activities. According to Brenner, "government surveillance is critical for identifying and disrupting terrorist plots before they come to fruition, making it an essential component of national security strategy."[1]

Surveillance technologies like wiretapping, data mining, and facial recognition are employed to monitor communications, track movements, and analyze behavior. These tools allow governments to gather vast amounts of data in real time, providing security agencies with the information they need to respond to emerging threats. However, the sheer scale of data collection raises concerns about overreach and the potential for abuse, particularly when surveillance programs are conducted without sufficient oversight or accountability.

The Erosion of Privacy Rights

One of the most significant concerns surrounding government surveillance is its impact on privacy rights. Civil liberties advocates argue that widespread surveillance undermines the right to privacy, a fundamental human right protected by international treaties such as the Universal Declaration of Human Rights and the European Convention on Human Rights. As surveillance technologies become more advanced, the potential for governments to collect, store, and analyze personal information without the individual's knowledge or consent has grown significantly.

The Snowden revelations in 2013 shed light on the extent of government surveillance programs, revealing that agencies like the NSA were engaged in mass data collection on a global scale. These disclosures sparked widespread outrage and fueled concerns about the erosion of privacy rights. According to Solove, "mass surveillance programs operate under the guise of national security, but in practice, they often infringe upon the privacy of ordinary citizens who are not involved in any criminal activity."[2]

Privacy advocates argue that surveillance programs should be subject to strict legal limits and oversight to prevent abuse and ensure that individuals' rights are protected. They contend that the collection of personal data without a warrant or due process violates the principles of a free society, where individuals are entitled to privacy and freedom from government intrusion. The challenge is finding a balance between the need for surveillance to protect national security and the need to uphold privacy rights.

Legal and Ethical Concerns Surrounding Mass Surveillance

The expansion of government surveillance has raised several legal and ethical concerns, particularly regarding the lack of transparency and accountability in many surveillance programs. Surveillance activities are often conducted in secrecy, with little public knowledge about the extent of data collection or how the information is used. This lack of transparency makes it difficult for citizens to hold governments accountable for potential abuses of power. However, informing individuals that their data is being collected undermines the effectiveness of government surveillance programs; it's akin to alerting potential threats, like terrorists, that they are being monitored.

One key legal concern is whether mass surveillance violates constitutional protections against unreasonable searches and seizures, as outlined in the Fourth Amendment of the US Constitution. While the courts have ruled that governments can collect certain types of data without a warrant, the scope of these rulings remains controversial. According to Bamford, "the expansion of surveillance powers has led to a legal gray area where governments can bypass traditional safeguards in the name of national security."[3]

Ethically, the issue of consent is also central to the debate. Individuals are often unaware that their data is being collected or analyzed, which raises questions about whether they have given meaningful consent to the intrusion. In many cases, individuals have little recourse to challenge surveillance activities, as the details of these programs are classified, and legal avenues for redress are limited.

The Chilling Effect on Free Speech and Civic Engagement

Another concern associated with government surveillance is its potential to chill free speech and civic engagement. When individuals know that their communications and online activities are being monitored, they may be less willing to express dissenting opinions, engage in political activism, or participate in public discourse. According to Penney, "mass surveillance has a chilling effect on freedom of expression, as individuals become reluctant to engage in activities that might be perceived as controversial or subversive."[4]

This chilling effect is particularly concerning in democratic societies, where the free exchange of ideas and open debate are essential for the functioning of a healthy political system. When surveillance is used to monitor and suppress dissent, it undermines the very foundations of democracy. For this reason, many civil liberties organizations, such as the American Civil Liberties Union (ACLU) and Human Rights Watch, have called for greater restrictions on government surveillance programs to protect free speech and political freedoms.

Oversight and Accountability in Government Surveillance

Many experts advocate for stronger oversight and accountability mechanisms to address concerns about government surveillance. Independent oversight bodies, such as parliamentary committees or judicial review panels, can help ensure that surveillance programs operate within the bounds of the law and respect civil liberties. These oversight bodies should have the authority to review surveillance activities, investigate potential abuses, and hold government agencies accountable for violations of privacy rights.

In the United States, the Foreign Intelligence Surveillance Court (FISC) plays a key role in authorizing and overseeing government surveillance activities related to national security. However, critics argue that the FISC operates in secrecy and rarely denies surveillance requests, raising concerns about the effectiveness of its oversight. According to Greenwald, "the FISC has been criticized for acting as a rubber stamp for government surveillance requests, with little meaningful scrutiny of the activities it authorizes."[5]

Greater transparency is also needed to ensure that the public is informed about the scope and nature of government surveillance. This includes making more information available about the data being collected, how it is used, and the safeguards in place to protect civil liberties. Transparency can help build public trust and ensure that surveillance programs are conducted in a manner consistent with democratic values.

Conclusion

The tension between government surveillance and civil liberties will likely remain a central issue in the digital age. While surveillance is an essential tool for protecting national security and preventing cyberattacks, it should be balanced with the protection of privacy rights and individual freedoms. Striking the right balance requires legal safeguards, ethical considerations, and robust oversight mechanisms to ensure that surveillance programs do not overreach or infringe on civil liberties. As surveillance technologies continue to evolve, governments should work to create transparent and accountable systems that protect both security and freedom.

It's worth noting that informing individuals of data collection significantly undermines the effectiveness of government surveillance programs; it's comparable to tipping off potential threats, such as terrorists, that they are under watch.

8.4 Corporate Responsibility in Data Breaches

In today's digital age, corporate responsibility in the event of data breaches is a critical issue. Companies hold vast amounts of sensitive personal data, including financial information, health records, and intellectual property. When a data breach occurs, the repercussions can be severe, affecting both the organization and the individuals whose data has been compromised. The ethical and legal implications of how corporations handle data breaches have become more prominent as the public increasingly expects transparency, accountability, and swift corrective action. Corporate responsibility extends beyond preventing breaches to include how companies respond when breaches occur, how they communicate with affected parties, and what steps they take to mitigate harm.

The Growing Frequency of Data Breaches

Data breaches have become common, affecting companies of all sizes and sectors. As digital infrastructures expand, so do the risks of cyberattacks and system vulnerabilities. The financial and reputational damage caused by data breaches can be devastating, especially for companies that fail to take appropriate security measures or respond effectively to breaches.

In recent years, high-profile breaches at companies such as Equifax, Target, and Yahoo have brought the issue of corporate responsibility to the forefront. These breaches exposed the personal data of millions of individuals and led to lawsuits, regulatory fines, and significant reputational damage for the organizations involved. For example, the Equifax breach in 2017, which compromised the personal information of 147 million people, resulted in a $700 million settlement with the US government.[1] This case highlights the high stakes for companies that fail to adequately protect their data.

Ethical Obligations of Corporations in Preventing Breaches

Corporations are ethically responsible for safeguarding the data they collect from customers, employees, and business partners. This responsibility includes implementing robust security measures to prevent unauthorized access to sensitive information. According to

Solove, "companies that collect and store personal data have an ethical obligation to protect it from breaches, as individuals entrust their private information to these organizations."[2]

Ethical responsibility also involves proactive risk management. This means conducting regular security audits, performing vulnerability assessments, and keeping up with the latest cybersecurity trends and threats. Negligence in these areas can lead to severe consequences, as failure to address known vulnerabilities can leave a company legally and ethically liable for the damages caused by a breach. Companies should view data protection as a core aspect of their operations rather than an afterthought.

Legal Responsibilities and Compliance with Data Protection Laws

In addition to ethical responsibilities, corporations have legal obligations to protect data, especially in jurisdictions with stringent data protection laws such as the European Union's GDPR and the California CCPA. These laws require companies to implement security measures to protect personal data and impose significant fines for non-compliance. Under GDPR, for example, companies can be fined up to 4% of their annual global revenue for failing to protect customer data[3].

Companies are also required by law to report data breaches to regulatory authorities and affected individuals within a specified time frame. Under GDPR, companies should report breaches within 72 hours of becoming aware of the incident, while the CCPA mandates that California residents be notified "in the most expedient time possible" after a breach. Failing to report breaches in a timely manner can result in additional legal penalties and further damage to a company's reputation.

One of the most high-profile cases of regulatory enforcement occurred in 2021 when British Airways was fined £20 million by the UK Information Commissioner's Office (ICO) for failing to protect the personal data of over 400,000 customers during a cyberattack in 2018.[4] The fine underscored the legal risks associated with inadequate data protection practices and reinforced the importance of compliance with data protection laws.

Corporate Responsibility in Post-Breach Responses

When a data breach occurs, how a company responds is just as important as the breach itself. Ethical corporate responsibility involves transparency, accountability, and a commitment to remediation. Companies should promptly inform affected individuals about the breach, provide details on what information was compromised, and outline the steps being taken to mitigate the damage. According to McKinsey, "companies that are transparent in their post-breach communications are more likely to retain customer trust and avoid long-term reputational damage." [5]

In addition to notifying affected parties, companies offer remedies, such as credit monitoring services or identity theft protection, to individuals whose data has been compromised. These actions demonstrate a commitment to mitigating harm and protecting individuals from the potential fallout of the breach. Furthermore, companies should conduct thorough investigations to determine the cause of the breach, implement security upgrades, and provide training to employees to prevent future incidents.

It's important to note that remedies like credit monitoring and identity theft protection can inadvertently increase individuals' vulnerability, as these services are often outsourced to third-party providers whose cybersecurity measures may be even weaker than those of the compromised companies.

The Role of Corporate Culture in Data Protection

Corporate culture plays a significant role in how companies handle data protection and respond to breaches. Companies that prioritize cybersecurity as part of their corporate ethos are more likely to implement effective security measures and foster an environment of vigilance. This includes creating cybersecurity awareness among employees, promoting the use of strong passwords and multi-factor authentication, and encouraging employees to report suspicious activities.

According to Deloitte, "Corporate leaders should take an active role in promoting a culture of security, as employees are often the first line of defense against cyberattacks."[6] When data protection is embedded in the corporate culture, employees are more likely to take security seriously and follow best practices. Conversely, companies that neglect to emphasize cybersecurity may find themselves more vulnerable to attacks and ill-prepared to respond to breaches.

Holding Companies Accountable for Data Breaches

Public and regulatory scrutiny of data breaches has increased in recent years as customers and government authorities hold companies accountable for failing to protect sensitive data. In many cases, companies that experience data breaches face class-action lawsuits, where affected individuals seek compensation for the damages they suffered. These lawsuits can result in significant financial penalties, as seen in the Yahoo data breach, which led to a $117.5 million settlement after the company was found to have mishandled a series of breaches affecting three billion user accounts. [7]

Regulatory authorities, such as the Federal Trade Commission (FTC) in the United States and the ICO in the United Kingdom, also play a critical role in holding companies accountable. These agencies investigate breaches, impose fines, and require companies to improve their data protection practices. The enforcement actions taken by regulators serve as a deterrent to other companies.

Conclusion

Corporate responsibility in data breaches is both an ethical and legal imperative in today's digital world. Companies have a duty to safeguard the data they collect, implement robust security measures, and respond transparently and responsibly when breaches occur. As the frequency and severity of data breaches continue to rise, organizations should take proactive steps to protect their systems and build a culture of cybersecurity. At the same time, regulatory authorities should continue to enforce data protection laws and hold companies accountable for failing to meet their obligations.

8.5 International Laws and Regulations on Cybersecurity

As cybersecurity threats become increasingly globalized, the need for international laws and regulations to govern cybersecurity has become a pressing issue. Cyberattacks know no borders and affect governments, businesses, and individuals worldwide. To combat these threats effectively, nations have implemented various cybersecurity laws and participated in international agreements aimed at harmonizing efforts across borders. However, despite these efforts, significant challenges remain in developing a unified global framework for cybersecurity. Countries often differ in their legal frameworks, regulatory approaches, and enforcement capabilities. This section explores key international cybersecurity laws, agreements, and challenges in the global regulation of cybersecurity.

The Need for International Cybersecurity Regulations

Cybersecurity threats are not limited to one country or region; they are global in scope and impact. Ransomware attacks, phishing, data breaches, and cyber espionage can disrupt international trade, national security, and critical infrastructure. As ENISA(the European Union Agency for Cybersecurity) notes, "cyber threats often originate from foreign actors, making international cooperation essential for addressing cross-border attacks and securing global digital infrastructure."[1] Individual countries may struggle to protect themselves from sophisticated cybercriminals or nation-state actors without coordinated international efforts.

The WannaCry ransomware attack in 2017, which affected over 150 countries, demonstrated the global nature of cybersecurity threats. The attack crippled healthcare systems, transportation networks, and businesses across the world. Such incidents highlight the need for a global regulatory framework that fosters cooperation and provides guidelines for responding to large-scale cyberattacks.

Key International Cybersecurity Laws and Agreements

Several key international laws and agreements have been developed to address cybersecurity threats. Among the most significant is the European Union's GDPR, which came into force in 2018. GDPR has set the standard for data protection and cybersecurity globally by imposing strict rules on how organizations handle personal data, mandating robust security measures, and requiring prompt breach notifications. According to the European Commission, "GDPR serves as a model for other countries seeking to protect personal data and secure information systems."[2]

Another major international initiative is the Budapest Convention on Cybercrime, which the Council of Europe adopted in 2001. This treaty, the first international agreement on cybercrime, provides a framework for harmonizing national cybersecurity laws, facilitating international cooperation, and promoting information-sharing between countries. The Budapest Convention aims to address crimes such as hacking, data theft, and online fraud, and it serves as a reference for countries developing their own cybercrime legislation. To date, over 65 countries have ratified or signed the treaty.[3]

Challenges in Harmonizing International Cybersecurity Laws

Despite these efforts, significant challenges remain in creating a unified global framework for cybersecurity. One major obstacle is the differing legal approaches taken by countries when regulating cybersecurity. For example, while GDPR focuses on protecting personal data and privacy, other countries, such as China and Russia, prioritize national security and control over internet activity. China's Cybersecurity Law, which came into force in 2017, imposes strict data localization requirements and gives the government broad powers to monitor and control internet use within the country.[4]

These differences in approach make it difficult to establish a common set of rules for cybersecurity across borders. As Schmitt argues, "the lack of consensus on how to balance privacy, security, and government control hampers efforts to create a cohesive international cybersecurity framework"[5]. Moreover, some countries may be reluctant to cooperate on international cybersecurity efforts due to concerns about national sovereignty and the potential for surveillance by foreign governments.

The Role of International Organizations in Cybersecurity

International organizations play a critical role in facilitating cooperation on cybersecurity issues. The United Nations (UN), through its Group of Governmental Experts (GGE), has sought to promote international dialogue on cybersecurity norms and responsible of member states in cyberspace. The GGE has issued reports emphasizing the need for countries to abide by international law in cyberspace, refrain from attacking critical infrastructure, and avoid engaging in cyber espionage.

Similarly, the International Telecommunication Union (ITU) has developed the Global Cybersecurity Agenda (GCA), which provides a framework for enhancing international cooperation on cybersecurity. The GCA focuses on five key areas: legal measures, technical and procedural measures, organizational structures, capacity building, and international cooperation. According to the ITU, "the GCA aims to create a secure and trustworthy global digital environment by fostering collaboration among governments, industry, and civil society."[6]

Regional Cybersecurity Initiatives

In addition to global initiatives, several regional cybersecurity agreements have been developed to address the specific needs of different regions. In the Asia-Pacific, the Asia-Pacific Economic Cooperation (APEC) has established the APEC Cybersecurity Strategy, which promotes cooperation on cybersecurity issues among its 21 member economies. The strategy focuses on building capacity, enhancing cybersecurity awareness, and facilitating cross-border collaboration to address cybercrime and secure the region's digital economy.

In Africa, the African Union (AU) adopted the Convention on Cyber Security and Personal Data Protection (commonly known as the Malabo Convention) in 2014. The Malabo Convention is the first regional treaty on cybersecurity and data protection in Africa, and it seeks to harmonize cybersecurity laws across the continent. The convention aims to strengthen national cybersecurity frameworks, protect personal data, and promote the adoption of best practices for securing information systems.[7]

The Role of Cyber Diplomacy

Cyber diplomacy has emerged as an important tool for addressing cybersecurity challenges at the international level. Countries engage in diplomatic efforts to establish norms for state behavior in cyberspace, negotiate international agreements, and foster collaboration on cybersecurity issues. According to Hathaway, "cyber diplomacy is essential for managing the complex relationships between states in the digital age, particularly as cybersecurity becomes a key component of national security strategies."[8]

For example, the US Department of State has established the Office of the Coordinator for Cyber Issues to lead diplomatic efforts on cybersecurity and promote international cooperation. The office works with foreign governments, international organizations, and the private sector to develop cybersecurity policies, combat cybercrime, and promote internet freedom. Similarly, the European Union has launched the EU Cyber Diplomacy Toolbox, which provides a framework for coordinating the EU's response to cyberattacks and promoting international norms in cyberspace [9].

Conclusion

The need for effective international laws and regulations to govern cybersecurity becomes even more pressing as cyber threats continue to change dynamically. While significant progress has been made through initiatives such as GDPR, the Budapest Convention, and regional agreements, challenges remain in harmonizing global cybersecurity efforts. Differing legal frameworks, national priorities, and concerns about sovereignty make it difficult to create a unified approach to cybersecurity regulation. However, the international community can work toward a more secure and resilient cyberspace through continued cyber diplomacy, international cooperation, and the development of global norms.

8.6 Ethical Considerations in AI-Driven Security Systems

Artificial intelligence (AI) has transformed the field of cybersecurity, offering tools that can rapidly detect, respond to, and prevent cyber threats. AI-driven security systems can analyze vast amounts of data in real time, identify patterns that indicate potential attacks, and automate responses to reduce human error. While these advancements offer significant benefits, they also raise important ethical considerations. The deployment of AI in security systems can lead to unintended consequences, including bias, privacy violations, and a lack of accountability. Addressing these ethical issues is crucial to ensuring that AI-driven security systems enhance cybersecurity without compromising fundamental rights.

The Promise of AI in Cybersecurity

AI has the potential to revolutionize cybersecurity by improving the speed and accuracy of threat detection. Traditional cybersecurity methods rely heavily on manual processes, which can be time-consuming and prone to human error. According to Gartner, "AI-driven security systems can analyze vast quantities of data faster and more accurately than human analysts, allowing organizations to detect threats in real-time."[1] AI's ability to process large datasets and identify anomalies makes it particularly effective in detecting sophisticated attacks, such as advanced persistent threats (APTs) and zero-day exploits.

AI-powered tools, such as machine learning (ML) algorithms, can be trained to recognize patterns in network traffic, identify suspicious behavior, and even predict future attacks. These capabilities allow organizations to move from reactive to proactive security, addressing vulnerabilities before cyber criminals exploit them. However, the very power of AI systems also raises ethical questions about their deployment, oversight, and potential for misuse.

Bias in AI Algorithms

One of the most pressing ethical concerns associated with AI-driven security systems is the potential for algorithmic bias. AI systems are trained on historical data, and if that data contains bias, the AI system may perpetuate and even exacerbate it. For example, suppose an AI-driven security system is trained on data that disproportionately identifies certain behaviors or user profiles as suspicious. In that case, it may lead to unfair targeting of specific groups. As noted by Binns, "bias in AI systems can result in discriminatory outcomes, particularly

279

when used in security contexts where fairness and impartiality are essential."[2]

Bias can have serious consequences in cybersecurity. If AI systems unfairly target specific individuals or groups, it can result in false positives, where innocent users are flagged as potential threats. This undermines trust in the technology and can lead to privacy violations and legal challenges. Addressing bias in AI algorithms requires careful consideration of the training data used, transparency in the decision-making process, and ongoing monitoring to ensure that the system is operating fairly.

Privacy and Surveillance Concerns

The use of AI in security systems often involves the collection and analysis of vast amounts of personal data. This raises significant privacy concerns, particularly when AI-driven security systems are used for mass surveillance. AI technologies, such as facial recognition and behavioral analysis, can be used to monitor individuals' movements, activities, and communications, potentially infringing on their privacy rights. According to Zuboff, "AI-driven surveillance systems represent a new form of power, one that threatens individual privacy and autonomy in unprecedented ways."[3]

In the context of cybersecurity, AI systems should strike a balance between the need for security and the protection of individual privacy. Organizations should ensure that the data collected by AI-driven security systems is used responsibly and that individuals' privacy is respected. This includes implementing data minimization practices, ensuring informed consent, and adhering to data protection regulations such as the GDPR. The ethical challenge is to prevent the misuse of AI technologies for unjustified surveillance while still leveraging their capabilities to enhance security.

Lack of Accountability and Transparency

Another major ethical issue in AI-driven security systems is the lack of accountability and transparency in decision-making processes. AI systems often operate as black boxes, where humans do not easily understand the reasoning behind their decisions. This lack of transparency can create challenges when AI systems make incorrect or biased decisions, as it becomes difficult to determine where the error occurred or who is responsible. According to Pasquale, "the opacity of AI systems makes it difficult to ensure accountability, particularly when decisions have significant consequences for individuals or organizations."[4]

In cybersecurity, this lack of transparency can be problematic when AI systems are used to make high-stakes decisions, such as blocking access to critical systems or flagging individuals as potential security threats. Without clear explanations for these decisions, affected parties may have little recourse to challenge them. This raises important ethical questions about the due process rights of individuals and organizations impacted by AI-driven security systems. To address these concerns, experts advocate for the development of explainable AI (XAI), which aims to make AI systems more transparent and their decisions more understandable.

Autonomy and the Risk of Over-Reliance on AI

AI-driven security systems offer significant advantages in terms of speed and efficiency, but there is a risk of over-reliance on AI at the expense of human judgment. Autonomous systems can act without human intervention, which raises concerns about the potential consequences of AI making critical decisions without adequate oversight. According to Rahwan, "Over-reliance on AI systems can lead to complacency, where human operators trust the technology too much and fail to exercise independent judgment."[5]

In cybersecurity, the consequences of over-reliance on AI could be severe. For example, an AI system might mistakenly block legitimate traffic or fail to detect a sophisticated cyberattack, leading to operational disruptions or security breaches. Human oversight remains essential to ensure that AI systems function correctly and that AI decisions align with ethical and legal standards. Organizations should implement safeguards to ensure that AI-driven security systems are used as tools to assist human decision-making rather than as a replacement for it.

Ethical Frameworks for AI in Cybersecurity

Several frameworks and guidelines have been developed to promote responsible AI deployment and address the ethical challenges posed by AI-driven security systems. For example, the European Commission's Ethics Guidelines for Trustworthy AI outline principles such as transparency, fairness, privacy, and accountability that should guide the development and use of AI technologies. [6] Similarly, the IEEE Global Initiative on Ethics of Autonomous and Intelligent Systems provides recommendations for ensuring that AI systems respect human rights and operate socially responsibly.

These frameworks emphasize the importance of developing AI systems that are fair, transparent, and accountable. They also call for regular audits of AI systems to ensure compliance with ethical standards and the involvement of diverse stakeholders in the design and deployment of AI technologies. By adhering to these principles, organizations can mitigate the ethical risks associated with AI-driven security systems and ensure that these technologies are used to enhance cybersecurity in an ethical and responsible manner.

Conclusion

The integration of AI into cybersecurity offers significant benefits, but it also raises complex ethical challenges. Issues such as bias, privacy violations, lack of accountability, and over-reliance on technology should be carefully considered to ensure that AI-driven security systems operate fairly, transparently, and responsibly. By adhering to ethical frameworks and implementing safeguards, organizations can harness the power of AI to improve security while protecting fundamental rights. There is a need for dialogue among technologists, ethicists, policymakers, and the public on how to navigate the ethical landscape of AI in cybersecurity.

Test Your Knowledge on Ethical and Legal Aspects of Cybersecurity

Q1: How do organizations balance privacy and security in cybersecurity?

Q2: What are the ethical considerations in ethical hacking and responsible disclosure?

Q3: How do government surveillance programs affect civil liberties?

Q4: What are the ethical concerns related to AI-driven security systems?

Q5: What role does corporate responsibility play in addressing data breaches?

Q6: How does the General Data Protection Regulation (GDPR) affect cybersecurity practices?

Q7: What are the legal ramifications of failing to comply with cybersecurity regulations?

Q8: How can companies ensure compliance with international cybersecurity laws?

Q9: How do ethical concerns influence the development of cybersecurity technologies?

Q10: What is the role of international agreements in improving cybersecurity?

Q11: How do organizations protect user data while complying with data protection laws?

Q12: What legal responsibilities do companies have when handling personal data?

Q13: How do governments collaborate with the private sector to enhance cybersecurity?

Q14: What are the ethical challenges of cyber warfare and nation-state hacking?

Q15: How does corporate transparency impact the aftermath of a data breach?

ANSWERS

Q1: A: Organizations should implement strong security measures while respecting privacy laws, ensuring that data protection does not infringe on individual rights.

Q2: A: Ethical hackers should obtain permission before testing systems and ensure that vulnerabilities are reported responsibly to prevent exploitation by malicious actors.

Q3: A: Government surveillance programs can infringe on civil liberties, such as the right to privacy, if not balanced with proper oversight and legal safeguards.

Q4: A: AI-driven security systems raise ethical concerns about bias, privacy, and the potential for misuse of surveillance technologies in ways that could harm individuals.

Q5: A: Corporations are responsible for protecting customer data and being transparent about breaches, ensuring timely disclosure and accountability in the event of a cyber incident.

Q6: A: GDPR mandates strict data protection measures and requires organizations to report breaches within 72 hours, increasing the emphasis on strong cybersecurity practices.

Q7: A: Non-compliance with cybersecurity regulations can result in hefty fines, legal penalties, and loss of business credibility, as seen in GDPR violations.

Q8: A: Companies should stay informed about relevant laws in each jurisdiction they operate in and implement robust cybersecurity measures that meet the highest regulatory standards.

Q9: A: Ethical concerns influence technology development by ensuring that security solutions respect human rights, such as privacy, and avoid discriminatory practices.

Q10: A: International agreements facilitate cooperation between nations in combating cybercrime, establishing shared frameworks for information sharing, enforcement, and incident response.

Q11: A: Organizations should implement encryption, access controls, and privacy policies that comply with laws like GDPR and HIPAA to protect user data.

Q12: A: Companies are legally required to protect personal data, report breaches, and ensure data processing is transparent, secure, and compliant with applicable laws.

Q13: A: Governments work with private companies through public-private partnerships, information-sharing initiatives, and joint defense strategies to enhance national cybersecurity.

Q14: A: Cyber warfare raises ethical challenges, as attacks on critical infrastructure can harm civilians, leading to concerns about the militarization of cyberspace and its impacts on global stability.

Q15: A: Corporate transparency in disclosing breaches and taking responsibility helps rebuild trust with customers, investors, and regulators, mitigating long-term reputational damage.

CHAPTER 9

GLOBAL CYBERSECURITY LANDSCAPE

Note: All references in this chapter are on page 404 - 405

This chapter turns the emphasis to the global side of cybersecurity, looking at how the modern threat picture is shaped by international cooperation, treaties, and cyberwarfare. Using cyber espionage, sabotage, and attacks on critical infrastructure will help you understand how nation-state actors get a tactical edge over rivals. With an eye on how governments, international organizations, and businesses should cooperate to share intelligence and coordinate defense measures, we will explore the value of worldwide cooperation in combating cybercrime. This chapter also examines international cybersecurity treaties and accords to establish criteria for responsible state action in cyberspace. Emphasizing how many countries vary in terms of their political systems, economic priorities, and technological capabilities, we will contrast and compare national cybersecurity policies. Since cybercrimes usually happen across borders and have different legal systems, the chapter ends with addressing the challenges related to cross-border data transfers and jurisdictional problems. By the end of this chapter, you will be completely aware of international rules, the situation of cybersecurity worldwide, and the need for coordinated defensive plans in the technologically linked world of today.

9.1 Cyber Warfare and Nation-State Activities

Cyber warfare has evolved into a natural component of contemporary conflicts in recent years. Nation-state players engage in offensive cyber actions to disrupt, destroy, or influence the operations of other nations' critical infrastructures. Unlike conventional warfare, cyber warfare takes place in the digital sphere where attackers take advantage of systems, infrastructure, and networks weaknesses.

These resources have become major objectives in geopolitics since governments, businesses, and people dependance more and more on digital technologies have rendered them indispensable. Often defined by their creativity, coordination, and accuracy, nation-state cyberattacks aim at causing extensive damage while preventing discovery. Developing reasonable defenses and rules to reduce the effects of cyberwarfare requires a grasp of its objectives, methods, and consequences.

The Rise of Cyber Warfare

The rise of cyber warfare is a direct result of the growing importance of cyberspace as a strategic domain. As noted by Clarke and Knake, "Cyber warfare is not merely a future possibility; it is already happening, with nation-states launching offensive cyber operations to achieve political, military, and economic objectives."[1] Cyber warfare has proven to be an effective tool for nations seeking to undermine their adversaries while avoiding the risks associated with conventional warfare. Unlike traditional military engagements, cyber warfare allows nation-states to strike at the heart of an enemy's critical infrastructure without crossing physical borders.

The Stuxnet attack in 2010 is often cited as the first major instance of nation-state cyber warfare. This highly sophisticated malware, believed to be developed by the United States and Israel, targeted Iran's nuclear enrichment facilities, causing significant damage to the country's nuclear program. According to Zetter, "Stuxnet demonstrated the potential for cyberattacks to inflict physical damage on critical infrastructure, marking a new era in the use of cyber capabilities in warfare."[2] Since then, numerous countries have developed and deployed cyber capabilities as part of their national defense strategies.

The Tactics and Techniques of Nation-State Cyberattacks

Nation-state cyberattacks often involve the use of advanced persistent threats (APTs), a long-term stealthy hacking campaigns designed to infiltrate and maintain access to targeted networks. These attacks are typically carried out by highly skilled actors who exploit zero-day vulnerabilities, use spear-phishing tactics, and leverage social engineering techniques to gain access to sensitive systems. Once inside, attackers can exfiltrate data, disrupt operations, or manipulate systems to cause damage.

A notable example of nation-state cyber warfare is the Russian cyberattacks against Ukraine, which escalated following Russia's annexation of Crimea in 2014. These attacks targeted Ukraine's power grid, government institutions, and financial systems, resulting in widespread blackouts and disruptions. In 2017, the NotPetya attack, which originated in Ukraine but spread globally, caused billions of dollars in damage by encrypting data and rendering systems inoperable.[3] The NotPetya attack demonstrated the potential for cyberattacks to have far-reaching effects, affecting companies and governments well beyond the intended target.

Cyber Espionage and Information Warfare

In addition to causing physical damage, nation-state cyberattacks are often used for cyber espionage and information warfare. Cyber espionage involves the theft of sensitive information, such as military plans, intellectual property, and government secrets, to gain strategic or economic advantages. China, for example, has been accused of engaging in cyber espionage activities targeting Western companies and governments to acquire valuable intellectual property and technology. According to Rid, "cyber espionage has become a central element of modern statecraft, with nations using cyber tools to gain insight into the capabilities and intentions of their rivals."[4]

Information warfare, on the other hand, involves the use of cyber tools to manipulate or disrupt the flow of information within a target country. This can include disinformation campaigns, hacking into media outlets, or manipulating social media platforms to influence public opinion and sow discord. The 2016 U.S. presidential election is a prominent example of how cyber tools can be used to interfere with the political process. Russian actors linked to the GRU were accused of hacking into political party servers and releasing sensitive information, as well as using social media to spread disinformation and influence voter behavior.[5]

The Attribution Challenge in Cyber Warfare

One of the key challenges in addressing cyber warfare is the difficulty of attribution. That is, determining who is responsible for an attack. Unlike conventional warfare, where the identity of the aggressor is usually clear, cyberattacks are often carried out by anonymous actors using proxy servers, encryption, and other techniques to obscure their origins. This makes it difficult for governments to definitively attribute attacks to specific nation-states and hold them accountable.

As noted by Rid and Buchanan, "the lack of clear attribution in cyber warfare creates a strategic dilemma for nations, as they may be hesitant to respond to attacks without concrete evidence of the perpetrator's identity."[6] The challenge of attribution complicates efforts to establish norms and rules of engagement in cyberspace, as states may deny involvement in cyberattacks or claim that non-state actors are responsible.

International Norms and Responses to Cyber Warfare

In response to the growing threat of cyber warfare, international organizations and governments have sought to develop norms and frameworks to regulate state behavior in cyberspace. The United Nations (UN) has taken steps to address this issue through its Group of Governmental Experts (GGE), which has called for states to refrain from attacking critical infrastructure and to respect international law in cyberspace. However, progress in developing binding international agreements on cyber warfare has been slow, as nations have differing views on how cyberspace should be governed.

In addition to diplomatic efforts, many countries are investing in cyber defense capabilities to protect themselves from nation-state attacks.

This includes the establishment of cyber commands within military structures, such as the U.S. Cyber Command, which is responsible for defending military networks and conducting offensive cyber operations when necessary. NATO has also recognized cyberspace as a domain of warfare, with the alliance committing to defend its members from significant cyberattacks.[7]

Conclusion

The role of cyber warfare in modern conflicts will continue to grow as nations increasingly rely on digital infrastructure for military, economic, and societal functions. Nation-states will likely continue to develop and deploy offensive cyber capabilities to achieve their geopolitical goals. The potential for cyberattacks to cause widespread damage remains a significant concern for global security. As Schmitt notes, "cyber warfare is likely to become a permanent fixture of statecraft, with nations seeking to exploit the vulnerabilities of their rivals through digital means."[8] Addressing the challenges of attribution, establishing international norms, and investing in cyber defenses will be critical to mitigating the risks posed by cyber warfare in the years to come.

9.2 International Cooperation in Fighting Cybercrime

As cybercrime continues to grow in both scale and sophistication, international cooperation has become essential in combating it. Cybercrime transcends national borders, with attacks originating in one country and often targeting individuals, businesses, and governments in another. The borderless nature of cybercrime makes it difficult for any single nation to combat these threats effectively on its own. This has led to the development of international agreements, joint law enforcement operations, and collaborative frameworks aimed at strengthening global cybersecurity efforts. However, the success of international cooperation in fighting cybercrime depends on the ability of countries to harmonize their laws, share information, and overcome geopolitical differences.

The Global Nature of Cybercrime

Cybercrime is a global issue that affects countries, industries, and individuals worldwide. Crimes such as ransomware, phishing, identity theft, and online fraud are perpetrated by criminal networks that operate across borders, making it difficult for national law enforcement agencies to track and apprehend perpetrators. According to Interpol, "cybercriminals exploit the lack of harmonization in international laws and jurisdictional challenges to evade justice."[1] This global dimension of cybercrime highlights the need for international cooperation to ensure that criminals cannot escape prosecution by exploiting legal gaps between countries.

An example of the global reach of cybercrime is the WannaCry ransomware attack in 2017, which affected over 200,000 computers across 150 countries, disrupting healthcare systems, businesses, and governments. The attack demonstrated the potential for cybercriminals to cause widespread damage on a global scale, underscoring the need for coordinated international responses to cybercrime.

Key International Cybercrime Treaties and Agreements

One of the most important international agreements in the fight against cybercrime is the Budapest Convention on Cybercrime, which was adopted by the Council of Europe in 2001. The convention serves as the first international treaty aimed at addressing crimes committed via the Internet and other computer networks. It provides a legal framework for the harmonization of national laws, the establishment of mutual assistance mechanisms, and the sharing of information between signatory states. To date, more than 65 countries have signed or ratified the convention, making it a key tool in global efforts to combat cybercrime.[2]

The Budapest Convention focuses on offenses such as illegal access, system interference, and data breaches, and it also promotes cooperation between law enforcement agencies to investigate and prosecute cybercriminals. According to ENISA, "the Budapest Convention has been instrumental in facilitating cross-border investigations and fostering cooperation between countries in tackling cybercrime."[3] However, the convention has faced criticism from countries such as Russia and China, which have expressed concerns about sovereignty and the influence of Western legal frameworks on their national laws.

In addition to the Budapest Convention, several regional initiatives have been developed to combat cybercrime. For example, the ASEAN region has implemented the ASEAN Cybersecurity Cooperation Strategy, which aims to promote collaboration among member states in fighting cybercrime. Similarly, the African Union (AU) adopted the Malabo Convention on Cybersecurity and Personal Data Protection in 2014, providing a framework for harmonizing cybersecurity laws across African nations.[4]

The Role of Interpol and Europol in Combating Cybercrime

Interpol and Europol play crucial roles in facilitating international cooperation in fighting cybercrime. Interpol's Global Cybercrime

Programme assists member countries in developing cybersecurity capabilities, conducting investigations, and sharing intelligence on cybercriminal networks. By providing a platform for information exchange and operational coordination, Interpol helps law enforcement agencies collaborate more effectively in tracking down cybercriminals across borders.

Similarly, Europol's European Cybercrime Centre (EC3) has been instrumental in coordinating joint law enforcement operations targeting cybercriminal groups operating in Europe and beyond. According to Europol, "the EC3 has successfully led several international operations that have dismantled major cybercriminal networks involved in activities such as ransomware, online fraud, and the distribution of illicit goods."[5]

One notable example of international cooperation facilitated by Europol is Operation GOLDFISH ALPHA, which took down a sophisticated cybercriminal group responsible for deploying ransomware across Europe. The operation involved law enforcement agencies from multiple countries, including the Netherlands, France, and the United States, and resulted in the arrest of key members of the criminal organization.[6] This demonstrates the effectiveness of international cooperation in disrupting cybercriminal activities that span multiple jurisdictions.

Challenges in International Cooperation

Despite the successes of international efforts to combat cybercrime, several challenges remain. One of the biggest obstacles is the lack of harmonization in national laws regarding cybercrime. Different countries have varying legal definitions of cyber offenses, and some lack comprehensive cybercrime legislation altogether. This creates difficulties in prosecuting cybercriminals who operate across borders, as laws that apply in one country may not be enforceable in another.

In addition to legal differences, geopolitical tensions can hinder international cooperation. Countries may be reluctant to share intelligence or collaborate on joint investigations if they perceive a rival nation as a potential cyber adversary. According to Schmitt, "geopolitical rivalries complicate efforts to establish global norms for cybersecurity, as nations may prioritize their own national interests over collective security." [7] The United States, China, and Russia have all been accused of engaging in cyber espionage and offensive cyber operations, creating mistrust that complicates international cooperation.

The Importance of Capacity Building

Another critical aspect of international cooperation in fighting cybercrime is capacity building. Many countries, particularly in the developing world, lack the resources and expertise to effectively combat cybercrime. International organizations and more developed nations play a key role in providing technical assistance, training, and support to help these countries build their cybersecurity capabilities.

For example, the United Nations Office on Drugs and Crime (UNODC)runs a Global Programme on Cybercrime, which provides training to law enforcement agencies in developing countries to enhance their ability to investigate and prosecute cybercriminals.[8] Similarly, Interpol offers training programs to help countries build their capacity to respond to cybercrime, with a focus on skills development, digital forensics, and incident response.

The Future of International Cooperation in Cybercrime

As cybercrime continues dynamically, so too should international cooperation in combating it. The future of international efforts to fight cybercrime will likely involve greater collaboration between governments, the private sector, and international organizations. Public-private partnerships are increasingly seen as essential for addressing cyber threats, as many of the tools and platforms used by cybercriminals are owned by private companies.

Moreover, the development of international norms and agreements governing state behavior in cyberspace will be crucial to ensuring that countries can work together effectively to combat cybercrime. According to Rid, "establishing norms for acceptable state behavior in cyberspace will help reduce tensions between nations and promote greater cooperation in fighting cybercrime."[9]

Conclusion

The fight against cybercrime requires a global response, as no single country can address these threats in isolation. International cooperation is essential for tracking and prosecuting cybercriminals who operate across borders. Despite challenges such as differing legal frameworks and geopolitical tensions, continued efforts to harmonize laws, share intelligence, and build capacity are vital for improving global cybersecurity. International collaboration will remain a cornerstone of efforts to protect individuals, businesses, and governments from the growing threat of cybercrime.

9.3 Global Cybersecurity Treaties and Agreements

As cyber threats continue to grow in complexity and impact, global cybersecurity treaties and international agreements have become increasingly vital for maintaining peace and security in cyberspace. These agreements provide frameworks for cross-border collaboration, harmonizing laws, and ensuring that nations adhere to common standards in addressing cybercrime, cyber espionage, and cyber warfare. Effective international cooperation, facilitated by treaties and agreements, is essential to combat cyber threats on a global scale.

The Need for Global Cybersecurity Agreements

The rise of global connectivity has made nations more vulnerable to cyberattacks. As noted by ENISA, "Cybersecurity is now a global issue, requiring coordination and cooperation among nations to secure the digital infrastructure that underpins our economies, societies, and governments."[1] While cybercriminals and state actors often exploit legal and jurisdictional gaps, international agreements help close these gaps by establishing common protocols for responding to cyber incidents and holding perpetrators accountable.

Without global cooperation, cyber threats could escalate, resulting in severe disruptions to critical infrastructure, economic systems, and national security. Treaties and agreements foster trust between nations, enabling the sharing of intelligence and resources necessary to prevent, mitigate, and respond to cyberattacks.

The Tallinn Manual and Cyber Warfare

One important document in the realm of global cybersecurity is the Tallinn Manual, which provides a comprehensive analysis of how existing international law applies to cyber warfare. Although not a legally binding treaty, the manual is a significant academic effort to clarify the legal norms surrounding state behavior in cyberspace during times of conflict. Developed by a group of international experts under the guidance of the NATO Cooperative Cyber Defence Centre of Excellence, the Tallinn Manual explores the application of international humanitarian law to cyber operations.

According to Schmitt, "the Tallinn Manual seeks to provide clarity on the legal frameworks that govern state conduct in cyber warfare, including the principles of distinction, proportionality, and necessity."[2] By outlining how traditional laws of war apply to cyberattacks, the manual helps states navigate the complex legal and ethical challenges posed by cyber conflicts.

Although the Tallinn Manual is not an official treaty, it has influenced discussions on the development of international norms for state behavior in cyberspace. The manual encourages states to consider the legal implications of their cyber operations and to respect the sovereignty and integrity of other nations' digital infrastructures.

The United Nations Efforts on Cybersecurity

The United Nations (UN) has also played a key role in promoting international cooperation on cybersecurity issues. The UN Group of Governmental Experts (GGE) has worked to establish norms and guidelines for state behavior in cyberspace. Since 2004, the GGE has issued several reports calling for states to refrain from attacking critical infrastructure, to respect international law in their cyber operations, and to cooperate in combating cybercrime.

The 2021 GGE report emphasized the importance of transparency, accountability, and confidence-building measures in cyberspace. According to the report, "states should establish mechanisms for communication and cooperation to prevent misunderstandings and reduce the risk of cyber conflicts escalating into full-scale warfare."[3] The GGE's work has contributed to the development of a common understanding of responsible state behavior in cyberspace, even as tensions between major cyber powers, such as the United States, China, and Russia, persist.

The UN has also supported efforts to develop regional cybersecurity agreements. For example, the Shanghai Cooperation Organisation (SCO), which includes China and Russia as key members, has developed its own framework for promoting cybersecurity and combating cyberterrorism. While these regional agreements differ in scope and emphasis from global initiatives like the Budapest Convention, they demonstrate the growing importance of international collaboration in addressing cyber threats.

Challenges in Developing Global Cybersecurity Treaties

Despite the progress made through treaties such as the Budapest Convention and the efforts of organizations like the UN, developing global cybersecurity agreements faces significant challenges. One major obstacle is the lack of consensus on how to balance security, privacy, and freedom of information in cyberspace. Countries have differing views on issues such as internet governance, data protection, and the role of the state in regulating online activity.

For example, while the European Union (EU) has prioritized data protection and privacy through regulations like the GDPR, other countries, such as China, emphasize state control over information and data localization. These differences make it difficult to develop a unified global approach to cybersecurity regulation.

Geopolitical tensions also complicate efforts to establish global cybersecurity treaties. Cyber espionage, cyber sabotage, and information warfare have become common tactics in international relations, with major powers often accusing each other of using cyber tools to gain political or economic advantages. These tensions create distrust and make it challenging to negotiate binding agreements that require mutual cooperation and transparency.

The Role of Public-Private Partnerships in Global Cybersecurity

In addition to treaties between governments, public-private partnerships (PPPs) are increasingly seen as essential for improving global cybersecurity. The private sector, which owns and operates much of the world's critical infrastructure, plays a crucial role in defending against cyberattacks. According to Deloitte, "public-private partnerships are key to addressing the challenges of cybersecurity at the global level, as governments and businesses should work together to protect critical systems and share threat intelligence."[4]

PPPs help bridge the gap between governments, which are responsible for creating regulatory frameworks, and private companies, which develop and maintain the technologies that power the digital economy. By sharing information about cyber threats and best practices, these partnerships enhance the ability of nations and businesses to respond to cyberattacks and minimize their impact.

Conclusion

The Budapest Convention, the Tallinn Manual, and the efforts of the UN represent important steps toward building a more secure and cooperative cyberspace. However, significant challenges remain, particularly in achieving consensus on key issues such as internet governance, data protection, and state sovereignty in cyberspace. To address these challenges, continued dialogue and collaboration among nations, international organizations, and the private sector will be essential. The future of global cybersecurity depends on the ability of nations to work together to develop frameworks that promote security, accountability, and respect for international law in the digital domain.

9.4 Comparative Analysis of National Cybersecurity Strategies

National cybersecurity strategies represent the framework through which countries address the growing challenges of cyber threats and the protection of critical infrastructure. Each nation, shaped by its unique economic, political, and technological environment, develops its own strategy to tackle cyber risks and ensure national security in the digital age. This comparative analysis examines the cybersecurity strategies of several leading nations, focusing on how different countries address common cybersecurity challenges such as cyber defense, information protection, and international cooperation. The analysis will explore the similarities, differences, and effectiveness of these strategies in addressing evolving cyber threats.

The United States: Cyber Defense and Public-Private Partnerships

The United States has developed one of the most comprehensive national cybersecurity strategies, emphasizing both cyber defense and public-private partnerships. The U.S. National Cyber Strategy places significant importance on defending critical infrastructure, protecting sensitive government data, and leveraging the private sector's expertise in cybersecurity. According to the White House Cybersecurity Strategy document, "the U.S. aims to strengthen its national defense posture by promoting collaboration between federal agencies, private companies, and international allies."[1]

A key element of the U.S. strategy is its focus on deterrence and offensive cyber capabilities. The strategy emphasizes the use of cyber deterrence to dissuade adversaries from launching attacks by maintaining the ability to launch counter-offensive operations if necessary. The U.S. has also established Cyber Command, a military unit tasked with defending national cyberspace and conducting offensive operations when required. The integration of cyber capabilities within the military highlights the importance the U.S. places on the role of cyber warfare in modern defense.

The U.S. strategy also stresses public-private partnerships, as the private sector owns and operates much of the nation's critical infrastructure. Cooperation between the government and private companies is seen as vital for improving cybersecurity resilience. These partnerships enable the sharing of threat intelligence, the development of best practices, and the deployment of cutting-edge technologies to protect against cyber threats.

The European Union: Data Protection and Privacy

In contrast to the U.S., the European Union's cybersecurity strategy places significant emphasis on data protection and privacy, driven by concerns over the misuse of personal data and the need to protect citizens' digital rights. The EU's General Data Protection Regulation

(GDPR) is central to its approach, setting strict rules for how businesses and governments collect, process, and protect personal data. According to ENISA, "the EU's strategy is rooted in a strong commitment to privacy and individual rights, balancing cybersecurity efforts with the protection of personal data."[2]

The EU's cybersecurity strategy also focuses on building resilience across its member states through cooperation and the creation of shared cybersecurity frameworks. The EU Cybersecurity Act, which was passed in 2019, established a Europe-wide certification framework to ensure that IT products and services meet cybersecurity standards. This approach reflects the EU's commitment to harmonizing cybersecurity policies across its member states, enhancing collective security while respecting national sovereignty.

Unlike the U.S., which emphasizes offensive cyber capabilities, the EU's strategy primarily focuses on defense, prevention, and resilience. The EU's approach reflects a preference for diplomatic solutions and the establishment of international norms in cyberspace. The EU has also been active in promoting international cooperation, seeking to develop global norms that govern responsible state behavior in cyberspace.

China: Cyber Sovereignty and State Control

China's cybersecurity strategy is distinct in its focus on cyber sovereignty and state control over the internet. The Chinese government views cyberspace as a domain that should be regulated and controlled to protect national security and promote the interests of the state. According to Deibert, "China's cybersecurity strategy emphasizes the concept of cyber sovereignty, which prioritizes the government's authority to regulate internet activities within its borders."[3] This approach has led to the development of strict data localization laws, which require foreign companies operating in China to store data on servers located within the country.

China's Cybersecurity Law, which came into effect in 2017, reflects the government's desire to maintain control over the digital economy and ensure that online activities align with national interests. The law requires companies to implement stringent cybersecurity measures, including regular security audits, encryption protocols, and data breach reporting requirements. While these measures are designed to improve cybersecurity, they also give the Chinese government extensive powers to monitor and regulate online activities.

China's cybersecurity strategy also includes a focus on technological self-reliance, with the government promoting domestic development of cybersecurity technologies to reduce dependence on foreign technology. This is part of China's broader goal of becoming a global leader in artificial intelligence (AI) and cybersecurity innovation.

Russia: Cyber Warfare and Information Control

Russia's cybersecurity strategy is heavily influenced by its emphasis on information control and cyber warfare. Russia views cyberspace as a critical domain for both defending national interests and exerting influence over other states. According to Rid, "Russia has been accused of using cyberattacks and information warfare to destabilize foreign governments, undermine democratic processes, and assert its geopolitical influence."[4]

Russia's strategy prioritizes offensive cyber capabilities, with cyber espionage and disinformation campaigns being key elements of its approach. Russian hackers have been linked to high-profile cyberattacks, including the 2016 U.S. election interference, Ukraine's power grid attack, and numerous cyber espionage operations targeting Western governments and organizations. These attacks demonstrate Russia's willingness to use cyber tools as part of its broader geopolitical strategy.

At the same time, Russia's cybersecurity strategy emphasizes information control within its own borders. The Russian government has enacted laws to regulate online content, monitor internet activities, and restrict access to foreign websites. These measures are part of a broader effort to control the flow of information and prevent external influence from undermining the government's authority.

Japan: Critical Infrastructure Protection and International Cooperation

Japan has developed a cybersecurity strategy that focuses on protecting critical infrastructure and enhancing international cooperation. Japan's reliance on technology and its advanced digital economy make cybersecurity a national priority, particularly in sectors such as energy, transportation, and finance. According to the Ministry of Internal Affairs and Communications, "Japan's cybersecurity strategy places significant emphasis on securing critical infrastructure to prevent disruptions that could have severe economic and social consequences."[5]

Japan's approach also emphasizes international cooperation, recognizing that cyber threats are a global issue that requires collaboration with other nations. Japan is an active participant in global cybersecurity initiatives, including those led by the United Nations, Interpol, and the Association of Southeast Asian Nations

(ASEAN). Japan has also sought to strengthen its cybersecurity partnerships with key allies, including the United States and Australia, to share intelligence, improve cybersecurity capabilities, and coordinate responses to cyberattacks.

Conclusion

While national cybersecurity strategies vary across countries, common themes such as the protection of critical infrastructure, the need for international cooperation, and the importance of public-private partnerships are prevalent. The United States emphasizes cyber defense and deterrence, while the European Union focuses on privacy and data protection. China and Russia prioritize state control and cyber sovereignty, viewing cyberspace as a domain for geopolitical maneuvering. Meanwhile, countries like Japan emphasize infrastructure protection and global cooperation.

Despite these differences, the global nature of cyber threats necessitates continued international collaboration. As nations develop their cybersecurity strategies, they should also consider the importance of building global norms, sharing intelligence, and coordinating responses to the increasingly complex and transnational cyber landscape.

9.5 Cross-Border Data Flow and Jurisdiction Issues

The globalization of the internet and the proliferation of digital services have led to unprecedented levels of cross-border data flow. This movement of data across international boundaries is crucial for the functioning of the global economy, powering everything from cloud services to e-commerce platforms. However, the growing volume of data that crosses borders has given rise to significant jurisdictional issues, as different countries have different laws and regulations governing the protection, storage, and transfer of data. These jurisdictional conflicts create challenges for governments, businesses, and individuals as they navigate the complex legal landscape of data sovereignty, privacy laws, and law enforcement access to data.

The Importance of Cross-Border Data Flow in the Global Economy

Cross-border data flow is essential for the smooth operation of the global economy. Digital services such as cloud computing, financial transactions, e-commerce, and social media rely on the seamless transfer of data between countries. According to McKinsey, "cross-border data flows contribute significantly to global economic growth, with data serving as the lifeblood of modern digital economies."[1] Many multinational companies store data in data centers located in different countries, enabling them to provide faster services to customers and optimize their operations globally.

The increasing reliance on cloud services has amplified the importance of cross-border data flow. Companies often store sensitive customer data in multiple locations around the world to ensure redundancy and resilience in the face of cyberattacks or natural disasters. This distributed approach to data storage has made it easier for businesses to scale globally, but it has also raised questions about the jurisdictional control of that data.

Jurisdictional Challenges and Conflicting Data Protection Laws

One of the most significant challenges associated with cross-border data flow is the issue of jurisdiction. When data flows across borders, it may be subject to the laws of multiple countries, creating conflicts between national data protection and privacy regulations. These jurisdictional conflicts often arise because different countries have different standards for how data should be handled, stored, and protected.

For example, the European Union's General Data Protection Regulation (GDPR) imposes strict rules on how personal data should be handled, even when that data is transferred outside of the EU. GDPR requires that companies transferring data to non-EU countries ensure that those countries provide an adequate level of protection for personal data. According to the European Commission, "the GDPR's rules on cross-border data transfers are designed to ensure that European citizens' personal data is protected, even when it is processed in third countries."[2]

However, countries such as the United States have less stringent data protection laws, which can lead to conflicts when data is transferred between jurisdictions with differing standards. This creates challenges for multinational companies that should navigate conflicting regulatory regimes to ensure compliance with both domestic and international laws. For example, the Privacy Shield Framework was designed to facilitate data transfers between the EU and the U.S., but it was invalidated by the European Court of Justice in 2020 due to concerns over U.S. government surveillance practices[3].

Data Sovereignty and Localization Requirements

The concept of data sovereignty refers to the idea that data is subject to the laws of the country in which it is located. Some countries, particularly those with strong concerns about national security or government surveillance, have implemented data localization requirements, which mandate that certain types of data should be stored within the country's borders. These laws are intended to ensure that governments retain control over sensitive data and can access it when needed for law enforcement or security purposes.

Countries such as China and Russia have implemented strict data localization laws, requiring companies to store data on local servers to ensure compliance with national regulations. According to Deibert, "data localization requirements are often justified on the grounds of national security, but they also serve as a tool for governments to exert greater control over the digital economy."[4] These laws create significant challenges for multinational companies that rely on the global free flow of data to operate efficiently.

Data localization requirements can also have a negative impact on innovation and the global economy. By forcing companies to store data locally, governments may inadvertently increase the costs of doing business and limit access to cutting-edge cloud services. Moreover, localization laws may fragment the internet, creating what some have called a "splinternet"—a division of the global internet into separate national or regional systems with differing rules and standards.

Law Enforcement Access to Cross-Border Data

Another major jurisdictional issue in cross-border data flow is the question of law enforcement access to data. As cybercrime becomes increasingly transnational, law enforcement agencies often need access to data stored in other countries to investigate crimes and bring perpetrators to justice. However, the process of obtaining data from foreign jurisdictions can be slow and complicated, often requiring mutual legal assistance treaties (MLATs) or other diplomatic agreements.

The United States' Clarifying Lawful Overseas Use of Data (CLOUD) Act, passed in 2018, addresses the issue of cross-border data access by allowing U.S. law enforcement to compel U.S.-based companies to provide data stored overseas, as long as the data is relevant to an ongoing investigation. The CLOUD Act also allows the U.S. to enter into executive agreements with foreign governments to facilitate cross-border data sharing for law enforcement purposes.[5]

While the CLOUD Act has been praised for improving the efficiency of cross-border data requests, it has also raised concerns about privacy and sovereignty. Some critics argue that the law allows governments to bypass domestic privacy protections by requesting data from companies based in foreign jurisdictions, potentially undermining the rights of individuals. According to Human Rights Watch, "the CLOUD Act could lead to violations of privacy and civil liberties if not implemented with strong safeguards."[6]

International Cooperation and the Future of Cross-Border Data Regulation

As the challenges surrounding cross-border data flow and jurisdiction continue to grow, international cooperation will be essential to finding solutions. Many experts argue that multilateral agreements are needed to harmonize data protection laws and establish clear rules for how data can be transferred and accessed across borders. The Budapest Convention on Cybercrime, which provides a framework for international cooperation in fighting cybercrime, is one example of how countries can work together to address these challenges.[7]

The development of global data protection frameworks, such as the OECD Guidelines on the Protection of Privacy and Transborder

Flows of Personal Data, is another step in the right direction. These guidelines promote the free flow of data across borders while ensuring that personal data is adequately protected. However, achieving global consensus on data protection standards remains difficult due to the differing priorities of individual countries.

Looking ahead, countries will need to balance their desire for data sovereignty with the need for global data flows that support innovation, economic growth, and the investigation of cybercrime. Public-private partnerships will also play a critical role in shaping the future of cross-border data regulation, as companies and governments collaborate to develop technologies and policies that protect privacy while enabling the efficient exchange of information.

Conclusion

Cross-border data flow is essential for the functioning of the global digital economy, but it also presents significant jurisdictional challenges. Conflicting data protection laws, data localization requirements, and law enforcement access to data stored in foreign jurisdictions complicate the legal landscape for businesses and governments. International cooperation and the development of harmonized regulatory frameworks will be crucial to addressing these challenges. By working together, countries can ensure that data flows freely across borders while respecting privacy, sovereignty, and security concerns

9.6 Building a Global Culture of Cybersecurity

As the digital world continues to expand and the number of cyber threats increases, it has become evident that building a global culture of cybersecurity is essential. Nations, businesses, and individuals should work together to develop practices, policies, and a shared understanding of cybersecurity's importance. Establishing a culture that prioritizes cybersecurity awareness, education, and collaboration across international borders is crucial to combating the evolving nature of cyber threats. This section explores the steps required to build a global cybersecurity culture and the role of international cooperation, public-private partnerships, and cybersecurity education.

The Need for a Global Cybersecurity Culture

The global interconnectedness brought by the internet and digital services has increased the need for a shared cybersecurity culture. With cyberattacks often affecting multiple countries simultaneously, a fragmented approach to cybersecurity only increases vulnerabilities. As stated by Schatz and Bashroush, "Cybersecurity is no longer an issue for individual organizations or nations; it is a global concern that requires a collective and coordinated effort."[1] A global cybersecurity culture emphasizes shared values, practices, and priorities that transcend borders, promoting a secure digital environment for everyone.

A successful global cybersecurity culture should address common cyber risks such as ransomware, phishing, data breaches, and nation-state attacks. However, achieving such a culture is challenging due to varying national policies, legal frameworks, and technological capabilities. The need for global cybersecurity norms and best practices is widely recognized as essential for harmonizing efforts.

The Role of International Cooperation

We have already dealt with International cooperation, Public-Private Partnerships, Interpol, Europol, and the United Nations efforts earlier and would not repeat them here.

The G20 and other international forums encourage member countries to work together on cybersecurity initiatives, including enhancing global resilience to cyberattacks and addressing cybercriminal activities. By fostering collaboration, nations can create a united front to deter malicious actors.

Internationally, organizations such as the World Economic Forum (WEF) have launched initiatives that encourage cooperation between governments, businesses, and civil society to address global cybersecurity challenges. For example, the WEF's Centre for Cybersecurity brings together stakeholders from around the world to collaborate on global solutions to combat cybercrime and secure the digital economy.

Cybersecurity Education and Awareness

Education and awareness are foundational to building a global cybersecurity culture. As cyber threats become more sophisticated, the need for a well-educated workforce capable of defending against these threats is greater than ever. According to Morgan and Bada, "cybersecurity education should be a global priority, aimed at equipping individuals and organizations with the knowledge and skills needed to protect themselves from cyber threats."[2]

Cybersecurity education should occur at multiple levels. First, at the individual level, raising awareness about safe online practices. Second,

314

at the organizational level, companies should invest in ongoing cybersecurity training for their employees. Finally, at the national level, governments should invest in developing a cybersecurity workforce capable of designing, implementing, and managing complex cybersecurity infrastructures.

International organizations are also playing a significant role in promoting cybersecurity education globally. For example, UNESCO and ITU have launched initiatives to enhance cybersecurity literacy, particularly in developing countries, by providing resources and training programs to help countries build their cybersecurity capabilities.

The Role of Cybersecurity Standards and Certifications

Establishing international cybersecurity standards is another essential step in building a global culture of cybersecurity. Standards provide a consistent framework for organizations to assess their cybersecurity posture and ensure that they are implementing best practices. According to ISO and IEC, "international standards for cybersecurity, such as the ISO/IEC 27001 framework, are critical for ensuring that organizations around the world adopt consistent and effective cybersecurity measures."[3]

Certifications for cybersecurity professionals, such as the Certified Information Systems Security Professional (CISSP), also play a significant role in ensuring that individuals possess the skills and knowledge necessary to protect organizations from cyber threats.

Adherence to international standards promotes trust between nations and organizations, facilitating cross-border data exchanges and cooperation. When organizations in different countries implement the same standards, it becomes easier to collaborate on cybersecurity initiatives and respond to incidents in a coordinated manner.

Challenges in Building a Global Cybersecurity Culture

Despite the many efforts to promote global cybersecurity, several challenges remain. One of the most significant is the digital divide between developed and developing countries. Many developing nations lack the infrastructure, resources, and expertise needed to implement robust cybersecurity measures. According to ITU, "the gap between developed and developing countries in terms of cybersecurity readiness remains wide, and addressing this disparity is essential for building a global cybersecurity culture."[4]

Additionally, geopolitical tensions between nations can hinder international cooperation on cybersecurity issues. Countries often accuse one another of cyber espionage, hacking, and other malicious activities, which can lead to mistrust and make collaboration difficult. For example, relations between the United States and China have been strained over allegations of state-sponsored hacking, complicating efforts to establish global cybersecurity norms.

Test Your Knowledge on Global Cybersecurity Landscape

Q1: What are the primary motivations behind nation-state cyber warfare?

Q2: How do international collaborations help combat cybercrime?

Q3: What are the key components of global cybersecurity treaties and agreements?

Q4: How do national cybersecurity strategies differ across countries?

Q5: What are the challenges of enforcing international cybersecurity laws?

Q6: How does cross-border data flow raise jurisdictional issues?

Q7: How do global cybersecurity standards improve organizational security?

Q8: How does cyber diplomacy contribute to global cybersecurity efforts?

Q9: How does the lack of cybersecurity infrastructure in developing countries impact global security?

Q10: What role do multilateral organizations play in global cybersecurity?

Q11: How do cyberattacks threaten international stability?

Q12: How do cultural and political differences affect global cybersecurity policies?

Q13: What is the role of cybersecurity norms in reducing cyber conflict?

Q14: How do international law enforcement agencies collaborate to fight cybercrime?

Q15: How can countries build a global culture of cybersecurity?

ANSWERS

Q1: A: Motivations include espionage, sabotage, economic advantage, and disrupting critical infrastructure, with the goal of gaining a strategic advantage.

Q2: A: International collaborations enable information sharing, joint investigations, and coordinated responses to cyber threats, making it harder for criminals to evade justice.

Q3: A: These agreements typically focus on information sharing, cyber defense cooperation, law enforcement collaboration, and establishing norms for responsible state behavior in cyberspace.

Q4: A: National strategies differ based on each country's economic priorities, political systems, threat landscape, and available resources for cybersecurity investment.

Q5: A: Challenges include jurisdictional issues, differing legal frameworks, and the anonymity of cyber criminals, which make it difficult to track and prosecute attackers across borders.

Q6: A: Cross-border data flow complicates enforcement because data stored or processed in different countries is subject to varying laws and regulations, creating conflicts over jurisdiction.

Q7: A: Global standards provide consistent guidelines for protecting digital assets, allowing organizations to align their cybersecurity practices with best practices across industries and regions.

Q8: A: Cyber diplomacy promotes international cooperation in addressing shared cybersecurity challenges, fostering dialogue between nations on norms, and conflict prevention in cyberspace.

Q9: A: Weak cybersecurity infrastructure in developing countries makes them more vulnerable to cyberattacks, which can have global consequences as interconnected systems are targeted.

Q10: A: Organizations such as the United Nations and NATO facilitate international cooperation, provide guidance on cybersecurity policies, and help coordinate responses to large-scale attacks.

Q11: A: Cyberattacks on critical infrastructure, such as energy grids or financial systems, can disrupt international stability by causing widespread damage and undermining trust between nations.

Q12: A: Differences in political and cultural values lead to varying approaches to privacy, security, and internet governance, complicating international collaboration on cybersecurity.

Q13: A: Establishing cybersecurity norms helps create expectations of responsible behavior in cyberspace, reducing the likelihood of cyber conflict and promoting cooperation between states.

Q14: A: Agencies collaborate through Interpol, Europol, and bilateral agreements, sharing intelligence, conducting joint operations, and pursuing cybercriminals across borders.

Q15: A: Countries can build a global culture of cybersecurity by promoting awareness, education, collaborative defense efforts, and adherence to international norms for digital security.

"In a world where data is the new currency, cybersecurity is not just a technical necessity, it is the foundation of trust and resilience in the digital age."

By Justin K. Kojok

CHAPTER 10

THE FUTURE OF CYBERSECURITY

Note: All references in this chapter are on page 406

This chapter investigates cybersecurity's prospects and challenges in the future. We will go over how quantum computing affects present encryption techniques. Quantum developments could call for totally fresh methods of data security. One of the main areas of emphasis will be AI-driven attacks and defenses. Security professionals and cybercriminals compete using artificial intelligence. You will also learn about edge computing and 5G network issues. These technologies provide better defense possibilities even while they create vulnerabilities. Data analytics is used in predictive cybersecurity to project and stop threats. Future cybersecurity success depends much on workforce development and education. By the end, you will appreciate the changing character of cybersecurity and the required competencies.

10.1 Quantum Computing: Threat and Opportunity

Quantum computing is set to revolutionize the computing world by offering vastly superior processing power compared to classical computers. However, while quantum computing presents significant opportunities for advancements in fields such as medicine, cryptography, and artificial intelligence (AI), it also poses serious cybersecurity threats. The ability of quantum computers to solve complex mathematical problems exponentially faster than classical computers could undermine many of the cryptographic techniques that currently protect sensitive data. At the same time, quantum computing offers opportunities for creating more secure encryption methods and enhancing cybersecurity defenses. This dual potential makes quantum computing both a threat and an opportunity in cybersecurity.

The Threat of Quantum Computing to Cryptography

One of the most significant threats posed by quantum computing is its potential to break widely used cryptographic systems, particularly those based on public-key encryption. Public-key encryption schemes, such as RSA and Elliptic Curve Cryptography (ECC), rely on the computational difficulty of factoring large numbers or solving discrete logarithm problems. Quantum computers, however, can use Shor's algorithm to solve these problems exponentially faster than classical computers, rendering these cryptographic methods obsolete. According to Michele Mosca, "quantum computing poses a major threat to the cryptographic foundations of today's Internet, as it can break the encryption systems that protect sensitive communications and financial transactions."[1]

The implications are far-reaching. If quantum computers become capable of breaking current encryption standards, sensitive data such as financial information, medical records, and government communications could be exposed to cybercriminals or nation-state actors. This would undermine the security of the Internet and many other digital systems that rely on encryption to protect data from unauthorized access. According to a study by NIST, "the advent of quantum computers could result in the compromise of all encrypted data, both at rest and in transit."[2]

The Timeline for Quantum Threats

While the threat of quantum computing to cybersecurity is real, experts believe that large-scale quantum computers capable of breaking modern encryption are still a decade or more away. While impressive in their capabilities, current quantum computers are not yet powerful enough to break RSA or ECC encryption. However, organizations cannot afford to be complacent. As noted by Chen and Jordan, "The development of quantum computers is progressing rapidly, and organizations should begin preparing for the quantum threat now to avoid being caught off guard"[3].

Some experts warn of the possibility of "harvest now, decrypt later" attacks, where attackers collect encrypted data today with the intention of decrypting it in the future when quantum computers become available. This is particularly concerning for finance, healthcare, and governments sectors, where sensitive data may need to remain secure for decades. As a result, organizations should start considering the implementation of quantum-resistant encryption techniques to safeguard their data in the long term.

Quantum-Resistant Cryptography: An Opportunity

Despite the significant threats posed by quantum computing, it also offers opportunities for enhancing cybersecurity through the development of quantum-resistant cryptography. Researchers are actively working on cryptographic algorithms that can withstand attacks from quantum computers. These algorithms, known as post-quantum cryptography (PQC), aim to provide security even in a future where quantum computers are widely available.

The National Institute of Standards and Technology (NIST)is leading efforts to standardize quantum-resistant cryptographic algorithms. In 2022, NIST announced the selection of four algorithms for further evaluation as part of its Post-Quantum Cryptography Standardization Project. According to NIST, "the goal of this project is to identify cryptographic algorithms that are resistant to quantum attacks and to develop standards that can be widely adopted across industries."[4]

One key advantage of PQC is that it can be implemented using existing hardware, making it easier for organizations to transition to quantum-resistant encryption. While it may take several years for these standards to be finalized and widely adopted, the development of PQC represents a significant opportunity to future-proof cybersecurity systems against the quantum threat.

Quantum Key Distribution (QKD): A New Paradigm for Secure Communication

In addition to quantum-resistant cryptography, quantum computing offers new opportunities for secure communication through Quantum Key Distribution (QKD). QKD is a method of generating and distributing cryptographic keys using the principles of quantum mechanics, which makes it virtually impossible for an eavesdropper to intercept the key without being detected. According to Bennett and Brassard, the pioneers of QKD, "quantum mechanics provides a fundamentally new way of ensuring the security of communication, as any attempt to intercept a quantum key will introduce detectable disturbances."[5]

QKD has the potential to revolutionize secure communication by providing unbreakable encryption. While current QKD systems are limited by their range and the need for specialized hardware, ongoing research is focused on overcoming these challenges to make QKD more practical for widespread use. Countries such as China have already made significant progress in developing QKD networks, with the launch of the world's first quantum communication satellite, Micius, in 2016. This satellite enables long-distance QKD, marking a significant step forward in the quest for quantum-secure communication.[6]

The Role of Governments and Industry in Preparing for Quantum Computing

Governments and industries worldwide are beginning to recognize the importance of preparing for the quantum computing revolution. Many governments, including those of the United States, China, and the European Union, have launched quantum initiatives to support research and development in quantum technologies, including quantum computing and quantum cryptography. According to the OECD, "governments should play a proactive role in supporting the development of quantum technologies and ensuring that industries are prepared for the quantum threat."[7]

The private sector is also investing heavily in quantum technologies, with companies such as Google, IBM, and Microsoft leading the charge in developing quantum computing hardware and software. These companies are not only focused on building more powerful quantum computers but are also working on solutions to mitigate the risks posed by quantum computing to cybersecurity. According to IBM Research, "quantum computing represents both a challenge and an opportunity for the cybersecurity industry, and companies should invest in developing quantum-safe solutions to protect their data."[8]

Conclusion

Quantum computing presents both threats and opportunities for the future of cybersecurity. While the ability of quantum computers to break current encryption methods poses a serious risk, the development of quantum-resistant cryptography and Quantum Key Distribution offers new possibilities for protecting sensitive data and securing communication. As quantum computing continues to advance, organizations, governments, and industries should take proactive steps to prepare for the quantum future. This includes investing in research, adopting quantum-resistant encryption, and collaborating on the development of global standards to ensure the security of the digital world in the post-quantum era.

10.2 AI-Powered Attacks and Defenses

The use of artificial intelligence (AI) in cybersecurity is a double-edged sword. On one hand, AI is being used to build more robust cybersecurity defenses by automating threat detection, analyzing patterns, and predicting potential vulnerabilities. On the other hand, cybercriminals are also using AI to launch more sophisticated and automated attacks that can bypass traditional security measures. The duality of AI in cybersecurity highlights its potential as both a threat and an opportunity. This section explores the implications of AI-powered cyberattacks and the defenses being developed to combat them.

The Rise of AI-Powered Cyberattacks

Cybercriminals increasingly use AI to launch more efficient and targeted attacks. AI can be trained to identify vulnerabilities in systems, automatically exploit them, and adjust its behavior based on the defenses it encounters. According to Goodfellow and Bengio, "AI enables attackers to automate complex attacks, allowing them to execute attacks at a scale and speed that would be impossible for human hackers."[1] This makes AI an invaluable tool for cybercriminals, especially in areas such as phishing, malware creation, and network infiltration.

One of the most concerning developments in AI-powered attacks is the rise of deepfake technology. Deepfakes use AI to create highly realistic fake images, videos, or audio recordings that can be used for malicious purposes. In a well-publicized case, cybercriminals used an AI-generated voice of a CEO to deceive a company executive into transferring $243,000 to a fraudulent account.[2] Such attacks, which rely on AI to manipulate human behavior, pose a significant threat to organizations and individuals alike.

AI is also being used to enhance the capabilities of traditional cyberattacks, such as ransomware and Distributed Denial of Service (DDoS) attacks. AI-powered ransomware can automatically adapt its attack strategy based on the target's defenses, making it more difficult to detect and mitigate. AI-driven DDoS attacks can dynamically adjust their behavior to avoid detection by traditional security tools and overwhelm networks with traffic while remaining hidden from defense systems.

AI-Enhanced Phishing Attacks

Phishing, a common form of cyberattack that involves tricking individuals into revealing sensitive information, has become more sophisticated with the help of AI. AI can generate convincing phishing emails or messages tailored to individual targets, making it more likely that the recipient will fall for the scam. According to Symantec, "AI-powered phishing attacks can analyze a target's online behavior and use that data to craft highly personalized and convincing messages."[3]

These AI-enhanced phishing attacks are often referred to as spear-phishing because they target specific individuals or organizations rather than relying on mass distribution. AI can analyze public data from social media profiles, corporate websites, and other online sources to create personalized phishing emails that are difficult to distinguish from legitimate communications. This increases the likelihood of success, as the recipient may be more inclined to trust a message that appears to be from a colleague or business partner.

Defending Against AI-Powered Attacks

Organizations are increasingly turning to AI-powered defenses to counter the growing threat of AI-powered cyberattacks. AI offers significant advantages in cybersecurity defense because it can process vast amounts of data in real time, identify patterns, and respond to threats faster than human analysts. AI-powered defense systems can also learn and adapt over time, improving their ability to detect and block new threats.

One of the key uses of AI in cybersecurity defense is in threat detection. Traditional threat detection systems rely on predefined rules and signatures to identify known threats. However, these systems struggle to detect new or evolving threats, such as zero-day exploits. AI-powered threat detection systems, on the other hand, can analyze network traffic, user behavior, and other data to identify anomalies that may indicate a cyberattack, even if the specific threat is not yet known.

Machine Learning and Behavioral Analysis

Machine learning (ML)is a subset of AI that enables systems to learn from data and improve their performance over time without being explicitly programmed. In cybersecurity, ML is used to analyze vast amounts of data and detect patterns that may indicate malicious activity. For example, ML algorithms can be used to analyze user behavior and detect deviations from the norm, which could indicate a compromised account or an insider threat. According to IBM Security, "machine learning is critical to identifying patterns in large datasets that human analysts might overlook, enabling faster detection of potential threats."[4]

By continuously learning from new data, ML-powered defense systems can adapt to evolving cyber threats. For example, User and Entity Behavior Analytics (UEBA) use ML to analyze user behavior and detect anomalies that could indicate insider threats or compromised credentials. This approach enables organizations to detect subtle indicators of cyberattacks that might otherwise go unnoticed.

AI in Incident Response and Automation

Another important application of AI in cybersecurity is in incident response. AI-powered systems can help automate the response to cyber incidents, reducing the time it takes to identify, contain, and mitigate a threat. According to McAfee, "automating incident response with AI allows organizations to respond to threats in real-time, minimizing the potential damage caused by a cyberattack."[5]

AI can be used to automatically quarantine compromised systems, block malicious traffic, or trigger alerts when suspicious activity is detected. By automating these tasks, organizations can reduce the time it takes to respond to a cyberattack and minimize the potential impact on their systems and data.

AI can also assist in forensic analysis following a cyber incident. After a breach has occurred, AI-powered tools can analyze logs, network traffic, and other data to determine the source of the attack, how it spread, and what vulnerabilities were exploited. This information can be used to prevent future attacks and improve security measures.

The Ethical Challenges of AI in Cybersecurity

While AI offers significant benefits in defending against cyberattacks, it also raises ethical challenges. One concern is the potential for AI to be used in autonomous cyberattacks, where AI systems are programmed to launch attacks without human intervention. Such attacks could cause widespread damage, particularly if they target critical infrastructure or other high-value assets. According to Brundage and Avin, "autonomous cyberattacks powered by AI could escalate conflicts in cyberspace, as attackers may deploy AI systems to launch attacks faster than defenders can respond."[6]

Another ethical issue is the potential for AI bias in cybersecurity systems. AI algorithms are only as good as the data they are trained on, and if that data contains biases, the resulting systems may make unfair or inaccurate decisions. For example, an AI-powered threat detection system may be more likely to flag certain types of traffic or users as suspicious based on biased training data. This could lead to false positives and the unfair targeting of individuals or groups.

Conclusion

AI has the potential to revolutionize cybersecurity, both as a tool for attackers and as a defense mechanism. While AI-powered attacks are becoming more sophisticated and dangerous, AI-powered defenses offer the ability to detect, mitigate, and respond to these threats more effectively than ever before. The continued development of AI in cybersecurity will require collaboration between governments, businesses, and researchers to ensure that AI is used ethically and responsibly.

10.3 Securing Emerging Technologies: 5G, Edge Computing, and Beyond

As emerging technologies such as 5G networks and Edge computing revolutionize industries and transform how data is processed and communicated, they also introduce new vulnerabilities and cybersecurity challenges. These technologies promise faster, more efficient, and decentralized communication systems, but their complexity and scale create new attack surfaces that should be secured. This section explores the cybersecurity challenges associated with 5G, Edge computing, and other emerging technologies, along with strategies to mitigate these risks.

The Promise and Perils of 5G Networks

5G technology represents the next generation of mobile network infrastructure, offering significantly faster speeds, lower latency, and the ability to connect more devices simultaneously than 4G. This advancement is critical for supporting the growing number of Internet of Things (IoT) devices, smart cities, and autonomous systems. However, the deployment of 5G also introduces unique cybersecurity risks. According to ENISA, "5G networks will exponentially increase the number of connected devices, which will, in turn, increase the potential attack surface for cybercriminals."[1]

The decentralized architecture of 5G networks, which relies on more small cell towers and distributed infrastructure, makes it more difficult to secure than traditional centralized networks. Choucri and Clark note that "5 G's distributed nature means that security measures need to be implemented at multiple layers, from the core network to the edge, creating new challenges for cybersecurity professionals."[2] Moreover, the integration of virtualization and software-defined networking (SDN) in 5G systems increases the complexity of securing these networks, as attackers may exploit vulnerabilities in the software used to manage and control the network infrastructure.

One of the most concerning security risks associated with 5G is the potential for nation-state actors to exploit the technology for espionage or sabotage. The involvement of foreign companies in building 5G infrastructure has raised concerns about the potential for backdoors that could be used for surveillance or cyberattacks. The controversy surrounding Huawei, a Chinese telecommunications company, and its role in building 5G networks in several countries highlights the geopolitical dimensions of 5G security.

Securing Edge Computing

Edge computing represents a paradigm shift in how data is processed, moving computation closer to the source of data generation rather than relying solely on centralized cloud servers. This shift reduces latency and improves efficiency, making edge computing essential for applications such as autonomous vehicles, smart manufacturing, and real-time analytics. However, it also introduces new cybersecurity challenges, as data is processed and stored at the network's edge, closer to end-user devices, and far from the traditional security measures of centralized data centers.

One of the key security risks in Edge computing is the distributed nature of the infrastructure, which increases the potential for data breaches and attacks on individual Edge nodes. As noted by Shi and Dustdar, "securing the edge requires a new approach to cybersecurity, as traditional perimeter-based security models are no longer sufficient to protect distributed systems."[3] Attackers can target vulnerable Edge devices, which may lack the robust security features of centralized data centers, to gain access to sensitive data or disrupt critical operations.

To address these risks, organizations should implement zero-trust security models that assume no device or user can be trusted by default. This approach requires continuous data monitoring, authentication, and data encryption at every stage of the data lifecycle. According to Forbes, "Edge computing environments should adopt a zero-trust approach to cybersecurity, ensuring that every device and user is authenticated and that all data is encrypted both at rest and in transit."[4] Additionally, organizations should implement artificial intelligence (AI) and machine learning (ML)to detect anomalies and respond to potential threats in real time.

IoT Security and Its Challenges

The integration of 5G and Edge computing is closely tied to the expansion of the Internet of Things (IoT), where billions of connected devices collect and transmit data. While IoT devices offer many benefits, they are often vulnerable to cyberattacks due to their limited computing power and lack of built-in security features. Attackers often target IoT devices such as smart home appliances, medical devices, and industrial sensors because they are easier to compromise than more heavily protected systems.

One of the major cybersecurity challenges associated with IoT is the difficulty in securing legacy devices that were not designed with security in mind. Many IoT devices have hardcoded passwords, unpatched vulnerabilities, or outdated firmware, making them attractive targets for cybercriminals. According to Weber, "IoT devices present unique security challenges because they are often deployed in environments where traditional security measures cannot be easily applied."[5]

To secure IoT ecosystems, device authentication, secure communication protocols, and automatic updates should be implemented to ensure that devices are protected from known vulnerabilities. Governments and industry bodies are also working on developing IoT security standards to ensure that devices meet minimum security requirements before they are deployed in the market. The European Union's Cybersecurity Act and the IoT Cybersecurity Improvement Act in the U.S. are examples of regulatory efforts aimed at improving IoT security standards.

The Role of AI and Machine Learning in Securing Emerging Technologies

As 5G, Edge computing, and IoT expand, AI and machine learning will play a critical role in securing these technologies. AI can analyze vast amounts of data generated by these systems in real time, identifying patterns, detecting anomalies, and predicting potential cyber threats. According to Gartner, "AI and machine learning are essential tools for securing emerging technologies, as they can process the massive amounts of data generated by 5G, IoT, and Edge computing environments faster and more accurately than human analysts."[6]

AI-powered threat detection systems can continuously monitor networks for signs of suspicious activity and respond to threats automatically. For example, behavioral analytics can be used to detect deviations from normal patterns in network traffic or user behavior, allowing organizations to identify and mitigate threats before they can cause significant damage. AI-driven automation can also improve the efficiency of incident response, reducing the time it takes to identify, contain, and resolve cyber incidents.

However, the use of AI in cybersecurity is not without its challenges. Attackers are also leveraging AI to create more sophisticated cyberattacks, such as AI-generated malware and automated phishing campaigns. This has led to an ongoing arms race between cybercriminals and cybersecurity professionals, as both sides use AI to gain an advantage.

Looking Beyond: Quantum Computing and Blockchain

While 5G, Edge computing, and IoT present immediate challenges, other emerging technologies, such as quantum computing and Blockchain, are also shaping the future of cybersecurity. Quantum computing, with its ability to break current encryption standards, poses a significant threat to cybersecurity, as discussed earlier. However, it also offers opportunities to develop quantum-resistant encryption and Quantum Key Distribution (QKD), which could enhance cybersecurity in the long term.

Blockchain technology also promises to improve cybersecurity by providing a decentralized and tamper-proof method for securing data. It can be used to secure IoT networks, ensuring the integrity of data transmitted between devices. According to Tapscott, "Blockchain's distributed ledger technology offers a new way to secure data in decentralized networks, making it a valuable tool for securing emerging technologies."[7]

Conclusion

The rapid adoption of 5G, Edge computing, and IoT offers significant opportunities for innovation and efficiency, but it also introduces new cybersecurity challenges. As these technologies become more integrated into everyday life, securing them will require a multi-layered approach that includes AI-driven defenses, zero-trust architectures, and robust security standards. Governments, industry leaders, and cybersecurity professionals should work together to address these challenges and ensure that emerging technologies are deployed safely and securely.

10.5 Predictive Cybersecurity and Proactive Threat Hunting

In the face of increasingly sophisticated cyber threats, predictive cybersecurity and proactive threat hunting have emerged as indispensable strategies for organizations seeking to stay ahead of attackers. Traditional reactive security measures, which focus on responding to threats after they occur, are no longer sufficient to combat modern cyberattacks. Predictive cybersecurity uses advanced analytics, artificial intelligence (AI), and machine learning (ML) to anticipate potential threats before they occur, while proactive threat hunting involves actively seeking out vulnerabilities and suspicious activities within a network before an attack occurs. Together, these approaches offer a more preemptive and dynamic defense against cyber threats.

The Shift from Reactive to Predictive Cybersecurity

Traditional cybersecurity approaches have relied heavily on signature-based detection, which identifies threats based on known patterns or behaviors. However, as cybercriminals develop more sophisticated attacks, including zero-day exploits and polymorphic malware, signature-based systems are proving inadequate. According to CrowdStrike, "reactive security models that focus on identifying known threats are no longer effective in today's rapidly evolving cyber threat landscape."[1] This has led to a shift toward predictive cybersecurity, which leverages AI and ML to analyze patterns, detect anomalies, and anticipate future attacks based on data-driven insights.

Predictive cybersecurity uses behavioral analysis, threat intelligence, and data analytics to build models that can forecast potential threats. By identifying patterns in network traffic or user behavior, predictive systems can detect deviations from the norm that may indicate malicious activity. According to Gartner, "predictive cybersecurity enables organizations to identify potential threats before they fully materialize, providing a critical advantage in preventing attacks."[2] This proactive approach allows organizations to implement preventive measures to mitigate risks rather than waiting for a breach to occur.

Artificial Intelligence and Machine Learning in Predictive Cybersecurity

The use of AI and ML is central to predictive cybersecurity. These technologies enable systems to learn from vast amounts of data and improve their accuracy over time, without the need for constant human intervention. According to Symantec, "AI-powered cybersecurity solutions can analyze millions of events in real-time, identifying suspicious behavior and predicting threats faster and more accurately than human analysts."[3]

Machine learning algorithms can be trained to recognize normal network behavior and detect deviations that may indicate a security breach. For example, unsupervised machine learning can analyze network traffic to identify outliers or anomalies that do not fit expected patterns, such as unusual login times or data transfers. Once these anomalies are detected, automated response systems can be triggered to block the suspicious activity or notify security personnel for further investigation.

Another important application of AI in predictive cybersecurity is the use of natural language processing (NLP) to analyze threat intelligence from multiple sources, including open-source data, dark web forums, and social media. NLP can extract valuable insights from unstructured data, such as identifying emerging cyberattack trends or spotting vulnerabilities in widely used software. By continuously analyzing and updating threat models, AI can help organizations stay one step ahead of cybercriminals.

Proactive Threat Hunting: Staying Ahead of Attackers

Proactive threat hunting is a critical component of modern cybersecurity strategies. Unlike traditional threat detection, threat hunting involves actively searching for signs of compromised systems, malicious activity, or vulnerabilities that have not yet been detected by automated tools. According to FireEye, "proactive threat hunting enables organizations to discover hidden threats that may be lurking within their networks, reducing the risk of undetected attacks."[4]

Threat hunting often relies on a combination of human expertise and automated tools. Skilled cybersecurity analysts use behavioral analysis, forensic techniques, and threat intelligence to identify potential threats and anomalies within the network. This approach is particularly effective in identifying advanced persistent threats (APTs), which involve long-term, targeted attacks by nation-state actors or sophisticated cybercriminal groups. APTs are designed to evade detection and maintain persistent access to a network over an extended period, making them difficult to detect using traditional security tools.

The key to successful threat hunting is the ability to detect indicators of compromise (IOCs)—signs that an attack may be in progress or that a system has already been compromised. IOCs can include unusual file activity, unexpected changes in system configurations, or irregular network traffic patterns. By identifying these indicators early, threat hunters can take steps to neutralize the threat before it escalates.

The Role of Threat Intelligence in Predictive Cybersecurity

Threat intelligence plays a crucial role in both predictive cybersecurity and proactive threat hunting. Threat intelligence refers to the collection and analysis of information about current and emerging cyber threats, including malware strains, attack methods, and the tactics, techniques, and procedures (TTPs) used by cybercriminals. According to Mandiant, "leveraging threat intelligence allows organizations to anticipate and defend against attacks by understanding the evolving tactics of threat actors."[5]

In predictive cybersecurity, threat intelligence feeds are used to continuously update threat models, ensuring that security systems are aware of the latest attack methods. For example, if a new type of malware is detected, predictive systems can quickly analyze the malware's behavior and update their defenses to block similar attacks. This real-time analysis enables organizations to stay ahead of cybercriminals who are constantly developing new attack techniques.

In proactive threat hunting, threat intelligence helps security analysts identify patterns of behavior that may indicate an attack in progress. For example, if a threat intelligence feed reports a surge in ransomware attacks targeting a specific industry, a security team may prioritize hunting for indicators of ransomware within their network. By using threat intelligence to guide their efforts, threat hunters can focus on the most relevant and pressing threats.

Challenges in Implementing Predictive and Proactive Cybersecurity

While predictive cybersecurity and proactive threat hunting offer significant advantages, there are also challenges associated with implementing these strategies. One of the primary challenges is the sheer volume of data that organizations should analyze to identify potential threats. Predictive cybersecurity systems should process vast amounts of network traffic, user behavior, and external threat intelligence, which can strain both computational resources and personnel.

Additionally, while AI and ML can enhance the accuracy of threat detection, they are not infallible. False positives—legitimate activities flagged as suspicious—can overwhelm security teams and lead to alert fatigue. According to IBM Security, "the high number of false positives generated by AI-powered security systems can cause significant operational challenges for organizations, leading to wasted time and resources."[6] To address this issue, organizations should continuously refine their AI models and ensure that human analysts are involved in reviewing critical alerts.

Another challenge is the complexity of threat landscapes. Cybercriminals are constantly evolving their tactics, making it difficult for security systems to keep up. Predictive cybersecurity systems should be able to adapt quickly to new and emerging threats, which requires continuous updates to threat models and security protocols.

Conclusion

As cyber threats become more sophisticated and frequent, predictive cybersecurity and proactive threat hunting will play an increasingly important role in defending against attacks. By leveraging AI, machine learning, and threat intelligence, organizations can anticipate and mitigate cyber threats before they cause significant harm.

Test Your Knowledge on The Future of Cybersecurity

Q1: How does quantum computing present both a threat and an opportunity for cybersecurity?

Q2: How can organizations prepare for the cybersecurity implications of quantum computing?

Q3: What is the role of AI-powered attacks in future cybersecurity challenges?

Q4: How can AI be used to defend against AI-powered cyberattacks?

Q5: What security challenges are associated with emerging technologies like 5G?

Q6: How does edge computing complicate cybersecurity strategies?

Q7: What are the potential security implications of self-driving cars and connected transportation systems?

Q8: How does predictive cybersecurity differ from traditional cybersecurity approaches?

Q9: What role does proactive threat hunting play in future cybersecurity strategies?

Q10: How can predictive threat intelligence help organizations stay ahead of cybercriminals?

Q11: What are the cybersecurity risks associated with autonomous systems?

Q12: How can organizations balance innovation with cybersecurity in deploying emerging technologies?

Q13: How might the rise of quantum computing impact cybersecurity for financial services?

Q14: What are the implications of biometric authentication in future cybersecurity?

Q15: How can organizations foster a culture of continuous learning in response to evolving cyber threats?

ANSWERS

Q1: A: Quantum computing poses a threat by potentially breaking traditional encryption methods, but it also offers opportunities for developing new, quantum-resistant encryption techniques.

Q2: A: Organizations can begin adopting post-quantum cryptography, invest in quantum research, and monitor advancements to ensure their encryption methods remain secure.

Q3: A: AI-powered attacks can automate the process of discovering vulnerabilities, launching sophisticated attacks, and evading detection, making cyber defense more complex.

Q4: A: AI can detect patterns, predict threats, and respond to attacks faster than humans, enabling organizations to defend against complex, AI-driven attacks in real time.

Q5: A: 5G increases the number of connected devices, expanding the attack surface and making monitoring and securing network traffic more difficult.

Q6: A: Edge computing decentralizes data processing, requiring new approaches to securing data at multiple endpoints, which can be more vulnerable to attacks.

Q7: A: Self-driving cars and connected systems could be vulnerable to **hacking**, potentially leading to accidents, service disruptions, or even targeted attacks on transportation infrastructure.

Q8: A: Predictive cybersecurity uses AI and machine learning to forecast potential threats before they occur, enabling organizations to strengthen defenses proactively.

Q9: A: Proactive threat hunting involves actively seeking out potential threats within a network rather than waiting for them to occur, improving an organization's ability to prevent attacks.

Q10: A: Predictive threat intelligence uses data analytics to anticipate future attack trends, enabling organizations to preemptively address vulnerabilities and avoid future attacks.

Q11: A: Autonomous systems, such as drones and AI-powered robots, can be hacked, misused, or manipulated to cause physical harm or disrupt operations.

Q12: A: Organizations can balance innovation and cybersecurity by adopting security-by-design principles, ensuring that security measures are integrated into new technologies from the outset.

Q13: A: Quantum computing could render traditional encryption used in financial transactions obsolete, requiring financial institutions to adopt new, quantum-safe encryption methods.

Q14: A: While biometrics provide enhanced security, they also present privacy concerns, and if biometric data is compromised, it cannot be changed like a password.

Q15: A: Organizations can encourage ongoing cybersecurity education, implement regular training, and promote awareness of emerging threats to ensure that employees remain vigilant and informed

CHAPTER 11

BUILDING CYBER RESILIENCE

Note: All references in this chapter are on page 407 - 408

Lastly, will go through the idea of building cyber resilience. That's how businesses get ready, defend and bounce back from cyberattacks. This chapter investigates practices of constant monitoring and adaptive security design. Strong defenses depend on ongoing improvement of security practices. You'll learn about collaborative defense tactics among different companies. There are public-private partnerships that let states and private companies work together to fight large-scale cyber threats. We will talk about their role. This chapter also shows you how to make and keep up-to-date incident reactions and business continuity plans, which will help businesses quickly recover from attacks.

11.1 Cyber Resilience

In today's world, cyber resilience has become a comprehensive approach that combines cybersecurity, business continuity, and organizational agility to ensure that a company can not only defend itself against cyber threats but also recover quickly and continue operations after an attack. It goes beyond traditional cybersecurity by focusing on an organization's ability to anticipate, withstand, recover from, and adapt to adverse cyber events. According to ENISA, "cyber resilience is the capability of an organization to ensure that its operations are not significantly impacted by cyberattacks."[1] The increasing complexity of cyberattacks and the interconnectedness of global digital infrastructure have made cyber resilience a critical component of modern business and government strategies.

This concept encompasses a range of strategies, including prevention, detection, response, and recovery. According to Cichonski and Franklin, "while cybersecurity focuses on protecting information systems from threats, cyber resilience emphasizes the ability to sustain essential operations even when those systems are compromised."[2] This broader approach recognizes that, in today's environment, it is not enough to prevent attacks; organizations should be prepared to recover quickly and minimize disruptions when they occur.

Cyber resilience is not limited to the IT department. It involves organizational processes, policies, training, and leadership to ensure an entity can operate under adverse conditions. A resilient organization understands that cyber incidents are inevitable and that the goal is to limit damage and restore operations as quickly as possible. As noted by Gartner, "The goal of cyber resilience is to ensure that, even in the face of an attack, the organization can continue to operate and deliver services to its customers."[3]

Measuring Cyber Resilience

Measuring cyber resilience is challenging due to the complex and evolving nature of cyber threats. Nonetheless, organizations should be able to assess their level of resilience to understand their readiness for potential cyber incidents. Several metrics and frameworks, ranging from quantitative indicators to qualitative assessments, are available to measure cyber resilience.

One widely used framework for assessing cyber resilience is the NIST Cybersecurity Framework (CSF), which provides a set of standards and best practices for managing and mitigating cybersecurity risks. The framework focuses on five core functions: Identify, Protect, Detect, Respond, and Recover. According to NIST, "measuring cyber resilience involves evaluating how well an organization performs across these five functions to ensure it can withstand and recover from cyber threats."[4]

Key Performance Indicator (KPIs), like the mean time to recovery (MTTR), incident response times, and the number of successful recoveries from hacks, can also be used by organizations to test their resilience. These measures give information about how well a company can handle incidents and keep downtime to a minimum. IBM Security says, "One of the most important metrics for measuring resilience is the speed and efficiency of an organization's response and recovery efforts."[5]

As part of cyber resilience assessments, organizations are often put through penetration tests, cyberattack simulations, and tabletop drills to see how well they can handle different types of cyberattacks. These exercises help companies become more resilient by finding holes in response protocols and recovery plans.

Key Components of Cyber Resilience

Several key components contribute to building cyber resilience in an organization:

1. **Preparedness and Prevention:** Preparedness includes developing incident response plans, conducting regular cybersecurity training, and ensuring all employees know their roles in responding to cyber incidents. According to Symantec, "proactive preparation and a strong culture of security awareness are critical to building cyber resilience."[6]

   Prevention involves implementing robust security controls, such as firewalls, intrusion detection systems, and encryption.

2. **Incident Response and Recovery: Incident response** is the process of detecting, responding to, and mitigating cyberattacks. A key element of cyber resilience is ensuring that organizations have well-defined response plans that can be executed swiftly in the event of an attack. Recovery involves restoring normal operations as quickly as possible while ensuring that critical business functions remain operational during the incident. Verizon emphasizes that "the speed at which an organization can recover from a cyberattack is a critical measure of its resilience."[7]

3. **Adaptability and Continuous Improvement:** Cyber threats are constantly evolving, and organizations should be able to adapt to new attack vectors. This requires continuous monitoring, regular updates to security systems, and ongoing training to ensure that employees are equipped to handle the latest threats. According to McAfee, "cyber resilience is not static; it requires

a continuous cycle of improvement, adapting to emerging threats, and refining security strategies."[8]

4. **Leadership and Governance:** Effective governance is crucial for building cyber resilience. Senior leadership should be involved in cyber risk management and ensure that the organization's resilience strategy aligns with its overall business objectives. Additionally, board members and executives should regularly review cyber risk reports and participate in cybersecurity decision-making. As Accenture notes, "Cyber resilience starts at the top, with leadership actively engaged in fostering a culture of security and resilience across the organization."[9]

Challenges in Achieving Cyber Resilience

While many organizations recognize the importance of cyber resilience, achieving it can be challenging due to several factors:

1. **Complexity of Cyber Threats:** The increasing complexity of cyberattacks, including ransomware, supply chain attacks, and advanced persistent threats (APTs), makes it difficult for organizations to stay ahead of adversaries. According to Kaspersky, "the evolving nature of cyber threats requires constant vigilance and the ability to respond to novel attack methods."[10]

2. **Resource Constraints:** Building and maintaining cyber resilience requires significant technology, personnel, and training investments. Small and medium-sized businesses (SMBs) often lack the resources to implement comprehensive resilience programs, leaving them vulnerable to attacks.

3. **Third-Party Risks:** Many organizations rely on third-party vendors for critical services, and these vendors can become entry points for cyberattacks. Ensuring that third-party partners have robust cybersecurity measures in place is essential for maintaining resilience. Deloitte highlights that "third-party risk management is a key component of building cyber resilience, as organizations should ensure that their vendors adhere to the same security standards."[11]

The Future of Cyber Resilience

The concept of cyber resilience has become increasingly important for organizations across all sectors. The COVID-19 pandemic has underscored the need for resilience, as organizations have had to quickly adapt to new threats and operational challenges, such as the rapid shift to remote work and the increased use of cloud services. According to Gartner, "the pandemic has accelerated the need for organizations to build resilience not just in their IT systems but across their entire business operations."[12]

Advances in artificial intelligence (AI) and machine learning (ML) will play a key role in enhancing cyber resilience by enabling organizations to detect and respond to threats more quickly. AI-driven solutions can analyze vast amounts of data in real time to identify anomalies and predict potential attacks, allowing organizations to respond proactively rather than reactively.

Conclusion

Building cyber resilience is essential for organizations seeking to navigate the complexities of the modern threat landscape. By focusing on preparedness, incident response, adaptability, and leadership, organizations can enhance their ability to withstand and recover from cyberattacks This ensures long-term success in an increasingly digital world.

11.2 Integrating Cybersecurity into Organizational DNA

In the rapidly evolving threat landscape, cybersecurity should be integrated into the very fabric of an organization's operations, culture, and decision-making processes. This concept of embedding cybersecurity into organizational DNA goes beyond implementing technical security controls; it involves building a security-conscious culture, aligning cybersecurity strategies with business objectives, and ensuring that every employee understands their role in protecting the organization's digital assets. By making cybersecurity a core element of organizational strategy and behavior, companies can better defend against cyber threats and enhance their overall cyber resilience. According to PwC, "organizations that embed cybersecurity into their DNA are more likely to respond effectively to cyber threats and maintain operational continuity during incidents."[1]

What It Means to Integrate Cybersecurity into Organizational DNA

Integrating cybersecurity into an organization's DNA means that cybersecurity awareness, policies, and practices are woven into every aspect of the organization, from daily operations to long-term strategic planning. It requires buy-in from senior leadership, employee engagement, and alignment with the organization's broader business goals. As noted by Gartner, "Cybersecurity cannot be treated as a siloed function; it should be embedded in the organizational mindset and considered in every decision."[2]

Cybersecurity should be a part of the organization's core values and mission, with clear policies and guidelines that reinforce the importance of security across all departments. This includes incorporating cybersecurity into employee training, risk management frameworks, product development, and even customer engagement. A security-conscious culture ensures that cybersecurity is not just the responsibility of the IT department but is shared across the entire organization.

The Role of Leadership in Cybersecurity Integration

Leadership plays a pivotal role in integrating cybersecurity into an organization's DNA. When executives and board members prioritize cybersecurity, it sets the tone for the rest of the organization. Leadership should demonstrate a commitment to security by providing adequate resources, budget, and support for cybersecurity initiatives. According to McKinsey, "cybersecurity needs to be a top priority for C-suite executives, and they should actively participate in shaping and guiding the organization's cybersecurity strategy."[3]

One effective way to ensure cybersecurity is integrated into leadership discussions is by appointing a Chief Information Security Officer (CISO) who reports directly to the CEO or the board of directors. The CISO's role is to advocate for cybersecurity at the highest levels, ensuring that cybersecurity considerations are factored into key business decisions. This alignment between cybersecurity and business strategy helps bridge the gap between technical teams and business leaders, ensuring that security risks are considered in all strategic decisions.

Similar should be applied in a governmental level. Cybersecurity and Infrastructure Security Agency (CISA) in the U. S. is an example.

Additionally, board members should receive regular cybersecurity briefings to stay informed about the organization's risk posture and emerging threats. According to Deloitte, "board members play a critical role in overseeing cybersecurity efforts, and their engagement in cybersecurity discussions can significantly enhance the organization's security posture."[4]

Building a Security-Conscious Culture

Creating a security-conscious culture is key to integrating cybersecurity into an organization's DNA. Employees at every level should be trained to recognize cyber risks, follow best practices, and understand their role in maintaining the organization's security. According to Verizon, "human error remains one of the leading causes of cyber incidents, making it essential for organizations to build a culture where security is second nature to all employees."[5]

Security awareness training should be mandatory for all employees and include education on topics such as phishing, social engineering, password hygiene, and secure data handling. These programs should be tailored to different roles within the organization, ensuring that employees in high-risk positions (e.g., those with access to sensitive information) receive more specialized training. Ongoing reinforcement of security awareness through regular drills, simulated attacks, and continuous education ensures that employees remain vigilant and informed about the latest threats.

In addition to training, organizations should encourage open communication about cybersecurity issues. Employees should feel comfortable reporting security incidents or potential vulnerabilities without fear of retaliation. A blame-free culture encourages transparency and fosters collaboration between employees and security teams.

Cybersecurity by Design in Product Development

Integrating cybersecurity into organizational DNA also means incorporating security-by-design principles into product development and service delivery. This approach ensures that cybersecurity is considered at the earliest stages of development rather than as an afterthought. According to NIST, "security-by-design involves embedding security features directly into the architecture of systems and applications to reduce vulnerabilities and enhance protection."[6]

For example, organizations should conduct security assessments and threat modeling to identify potential risks early in the design phase when developing new software or launching a digital service. These assessments help prevent security gaps and ensure that proper encryption, access controls, and data protection measures are built into the product. Adopting security-by-design principles not only reduces the likelihood of future breaches but also demonstrates to customers and stakeholders that the organization takes cybersecurity seriously.

Aligning Cybersecurity with Business Objectives

A critical aspect of integrating cybersecurity into organizational DNA is ensuring that security efforts are aligned with the organization's business objectives. Cybersecurity should be seen as a strategic enabler that supports the organization's ability to achieve its goals rather than a cost center. According to Accenture, "when cybersecurity is aligned with business priorities, it becomes a value driver that enhances trust, customer loyalty, and competitive advantage."[7]

For instance, organizations in highly regulated industries, such as finance and healthcare, should align their cybersecurity efforts with compliance requirements to protect sensitive data and avoid regulatory penalties. Similarly, e-commerce companies should prioritize data privacy and secure payment processing to build customer trust and protect their reputation. Organizations can ensure that their security strategies support growth, innovation, and customer satisfaction by aligning cybersecurity with business goals.

Challenges in Integrating Cybersecurity

Despite the benefits of integrating cybersecurity into organizational DNA, organizations may face several challenges. One major challenge is the fragmented approach to cybersecurity, where different departments or teams may have their own security practices or technologies that are not aligned with the organization's overall strategy. This can create gaps in security and increase the organization's exposure to risk. According to Symantec, "a siloed approach to cybersecurity often leads to inefficiencies, duplication of effort, and increased vulnerability to attacks."[8]

Additionally, resource constraints can make it difficult for organizations to invest in the necessary tools, technologies, and personnel required to build a robust cybersecurity program. Small and medium-sized businesses (SMBs) may struggle to allocate sufficient budget to cybersecurity, which can leave them vulnerable to cyberattacks. To address these challenges, organizations should prioritize cybersecurity investments and ensure that security is viewed as a strategic imperative at all levels of the organization.

Conclusion

Integrating cybersecurity into organizational DNA is essential for building long-term cyber resilience and ensuring that organizations are prepared to defend against the ever-growing array of cyber threats. By embedding security into leadership, culture, product development, and business strategy, organizations can create a holistic approach to cybersecurity that permeates every aspect of their operations.

As cyberattacks become more sophisticated, organizations should move beyond a reactive approach to cybersecurity and adopt proactive measures that involve all employees and departments. By building a security-conscious culture, aligning cybersecurity with business objectives, and ensuring leadership involvement, organizations can better protect their digital assets, maintain operational continuity, and foster trust with their customers and stakeholders.

11.3 Adaptive Security Architecture

Organizations should adopt an adaptive security architecture in the current cybersecurity landscape, where threats are constantly evolving and becoming more sophisticated. Unlike traditional security models that rely on static defenses, adaptive security is dynamic and continuously adjusts to new threats and vulnerabilities. An adaptive security architecture integrates advanced technologies such as artificial intelligence (AI), machine learning (ML), and behavioral analytics to anticipate, detect, respond to, and recover from cyberattacks in real time. As defined by Gartner, "an adaptive security architecture continuously assesses and adapts to evolving risks, providing organizations with a more flexible and responsive defense against cyber threats."1

What Is Adaptive Security Architecture?

Adaptive security is a proactive approach that focuses on the entire lifecycle of a cyberattack, from detection to recovery. Traditional security models are primarily preventive in nature, relying on perimeter defenses such as firewalls and antivirus software to block known threats. However, these defenses are often insufficient against zero-day attacks and advanced persistent threats (APTs), which can bypass traditional security measures. According to Symantec, "static security solutions are no longer effective in today's dynamic threat environment, making adaptive security a necessary evolution in cybersecurity strategy."2

Adaptive security, by contrast, is dynamic and continuous, incorporating real-time monitoring, threat intelligence, and automated responses. The architecture is built on four key pillars: predictive, preventive, detective, and responsive capabilities. This framework allows organizations to not only defend against known threats but also adapt to new attack vectors as they emerge. According to Cisco, "adaptive security combines predictive analytics with real-time monitoring to provide an integrated approach to threat detection and response."3

Predictive Security: Anticipating Threats

The predictive component of an adaptive security architecture involves using data analytics and AI to anticipate potential threats before they occur. By analyzing historical data, network traffic, and user behavior, predictive systems can identify patterns that may indicate an imminent attack. For example, AI can analyze anomalies in network traffic, such as unusual login times or unexpected file transfers, to predict potential breaches. According to CrowdStrike, "predictive analytics play a crucial role in identifying and mitigating risks before they manifest as full-blown attacks."[4]

In addition to identifying suspicious patterns, predictive security systems can also leverage threat intelligence feeds to stay updated on the latest attack methods and vulnerabilities. This allows organizations to anticipate new attack techniques and proactively implement countermeasures. For instance, if threat intelligence indicates a rise in ransomware attacks targeting a specific industry, predictive systems can adjust security policies to strengthen defenses against this threat.

Preventive Security: Building Strong Defenses

Preventive security is the traditional aspect of cybersecurity that focuses on blocking threats before they can penetrate the network. In an adaptive security architecture, preventive measures are continuously updated based on the latest threat intelligence and real-time analytics. This includes implementing firewalls, intrusion prevention systems (IPS), and endpoint protection that can adapt to new threats as they are identified.

The key difference between preventive security in an adaptive architecture and traditional approaches is that adaptive security systems are self-learning and automated. This means they can adjust their defenses based on the evolving threat landscape without the need for manual intervention. According to Palo Alto Networks, "an adaptive security architecture ensures that preventive controls are constantly updated and fine-tuned to address emerging threats in real time."[5]

Detective Security: Monitoring and Identifying Attacks

The detective layer of an adaptive security architecture involves continuous monitoring to detect potential breaches as soon as they occur. Traditional security solutions often fail to detect attacks until significant damage has been done, but adaptive security systems use real-time analytics and AI-driven detection to identify attacks in progress. According to Forrester, "real-time detection is a critical element of adaptive security, allowing organizations to identify and contain breaches before they escalate."[6]

Behavioral analytics is a key tool in the detective layer, as it enables systems to detect anomalies in user behavior that may indicate a compromised account. For example, suppose a user typically logs in from a specific location during business hours but suddenly logs in from a different location at an unusual time. In that case, the system will flag this as suspicious activity. By continuously learning what constitutes normal behavior, adaptive security systems can detect even subtle deviations that may indicate a breach.

Responsive Security: Automated Incident Response

The responsive component of an adaptive security architecture focuses on automated incident response to minimize the impact of a breach. Once an attack is detected, the system automatically takes action to contain the threat, such as isolating compromised systems, blocking malicious traffic, or initiating a rollback of affected files. According to IBM Security, "the ability to respond automatically to threats is one of the key advantages of adaptive security, as it allows organizations to limit damage and recover quickly."[7]

In addition to automated responses, adaptive security systems also provide forensic analysis to determine the root cause of an attack and ensure that similar incidents do not happen again. This continuous detection, response, and improvement cycle enables organizations to become more resilient to cyber threats over time.

The Role of Artificial Intelligence in Adaptive Security

Artificial intelligence is central to the effectiveness of adaptive security architectures. AI-powered systems can process vast amounts of data in real time, identifying patterns and anomalies that human analysts might miss. According to McAfee, "AI is the driving force behind adaptive security, enabling organizations to stay ahead of cybercriminals by predicting and responding to threats faster than ever before."[8]

Machine learning (ML) algorithms can be trained to recognize normal network behavior and detect deviations that may indicate malicious activity. Over time, these algorithms become more accurate as they learn from new data and improve their ability to detect emerging threats. AI-driven automation also plays a key role in reducing the time it takes to respond to incidents, as systems can take immediate action to contain and mitigate attacks without waiting for human intervention.

Challenges in Implementing Adaptive Security Architecture

Despite its many benefits, implementing an adaptive security architecture comes with challenges. One of the primary challenges is the complexity of integrating advanced technologies such as AI, machine learning, and behavioral analytics into existing security infrastructures. According to Gartner, "organizations often struggle to integrate adaptive security technologies due to the complexity of legacy systems and the lack of skilled personnel."[9]

Additionally, cost can be a barrier, particularly for small and medium-sized businesses (SMBs). Adaptive security systems often require significant hardware, software, and training investment. However, many experts argue that the long-term benefits of improved detection and faster response times outweigh the initial costs.

Conclusion

The sophisticated nature of cyber threats calls for organizations to move beyond traditional static security models and adopt adaptive security architectures that can respond to the evolving threat landscape. By integrating predictive, preventive, detective, and responsive capabilities, adaptive security provides a more flexible, real-time defense that allows organizations to anticipate and mitigate attacks before they cause significant damage.

While implementing adaptive security requires investment in advanced technologies and skilled personnel, the benefits in terms of improved threat detection, faster incident response, and increased resilience are undeniable. As noted by Gartner, "Adaptive security is the future of cyber defense, enabling organizations to stay one step ahead of attackers and minimize the impact of cyber incidents."[10]

11.4 Continuous Monitoring and Improvement

Continuous monitoring and improvement is a critical concept in modern cybersecurity, emphasizing the need for organizations to maintain an ongoing, proactive approach to securing their digital assets. As cyber threats grow more sophisticated and persistent, relying on static security controls or periodic assessments is insufficient to protect against evolving risks. Continuous monitoring enables organizations to detect, respond to, and mitigate threats in real time, while continuous improvement ensures that security measures evolve in response to new vulnerabilities and emerging attack techniques. According to NIST, "continuous monitoring is essential for maintaining situational awareness, assessing security controls, and ensuring that risks are effectively managed over time."[1]

The Importance of Continuous Monitoring

Continuous monitoring is the process of collecting, analyzing, and responding to security-related information on an ongoing basis to detect potential threats or vulnerabilities. It provides organizations with real-time insights into the health of their security systems and allows them to identify anomalous activities that may indicate a cyberattack. Unlike periodic assessments, which only provide a snapshot of an organization's security posture, continuous monitoring ensures that organizations have up-to-the-minute visibility into their environments.

One of the key advantages of continuous monitoring is its ability to detect threats that traditional security tools might miss. For example, many advanced cyberattacks, such as advanced persistent threats (APTs), are designed to evade detection by staying hidden within the network for extended periods. Continuous monitoring can detect subtle anomalies that may indicate the presence of an APT or other sophisticated threats by continuously analyzing network traffic, user behavior, and system configurations. According to Forrester, "Organizations that implement continuous monitoring are better equipped to detect and respond to advanced threats, reducing the time to identify and contain breaches."[2]

Key Components of Continuous Monitoring

A comprehensive continuous monitoring strategy involves several key components:

1. **Real-Time Threat Detection:** Continuous monitoring relies on real-time analytics to detect potential threats as they occur. Organizations can quickly identify deviations from normal patterns that may indicate an attack by monitoring network traffic, system logs, and user behavior. According to Gartner, "real-time threat detection enables organizations to respond to incidents in minutes rather than days or weeks, reducing the impact of cyberattacks."[3]

2. **Automated Alerts and Response:** Automated alerts are a critical aspect of continuous monitoring, as they allow security teams to respond to incidents quickly and efficiently. When the system detects an anomaly or potential threat, it can trigger an alert that notifies security personnel or initiates an automated response, such as blocking malicious traffic or isolating compromised systems. IBM Security notes that "automated response capabilities are essential for minimizing the damage caused by cyberattacks, as they allow organizations to take immediate action before a threat escalates."[4]

3. **Continuous Risk Assessment:** Continuous monitoring also involves ongoing risk assessment to evaluate the effectiveness of security controls and identify new vulnerabilities. This includes regularly scanning for security weaknesses, misconfigurations,

and outdated software that attackers could exploit. Organizations can prioritize patch management and system updates by continuously assessing risks to mitigate vulnerabilities before they are exploited. According to NIST, "continuous risk assessment is essential for ensuring that security controls remain effective and aligned with evolving threats."[5]

The Role of Continuous Improvement

In addition to monitoring, continuous improvement is critical to an organization's cyber resilience strategy. Continuous improvement involves regularly reviewing and refining security processes, tools, and policies to ensure that they remain effective in the face of emerging threats. This requires a cycle of assessment, response, and refinement that enables organizations to adapt to new attack methods and vulnerabilities.

One of the primary drivers of continuous improvement is the lessons learned from past incidents. After a cyberattack or security breach, organizations should conduct a post-incident review to identify what went wrong, how the attack was detected, and how it could have been prevented. According to McAfee, "continuous improvement is built on the principle of learning from incidents and implementing changes that make future breaches less likely or less impactful."[6]

Cybersecurity Maturity Models

To facilitate continuous improvement, many organizations use cybersecurity maturity models to assess their security posture and identify areas for growth. These models provide a structured framework for evaluating security practices and measuring progress over time. One widely used model is the Cybersecurity Maturity Model Certification (CMMC), which provides organizations with a roadmap for improving their cybersecurity capabilities. According to Deloitte, "maturity models help organizations benchmark their current security practices against industry standards and identify specific areas for improvement."[7]

Maturity models typically categorize cybersecurity capabilities into several levels, ranging from basic protections to advanced, adaptive defenses. By progressing through these levels, organizations can build a more resilient security posture capable of withstanding known and unknown threats.

Challenges of Implementing Continuous Monitoring and Improvement

While continuous monitoring and improvement are essential for cybersecurity, implementing these strategies is not without challenges. One of the primary challenges is the volume of data generated by continuous monitoring tools. Security teams should be able to process and analyze large amounts of data in real time to identify threats, which can strain both technological resources and human analysts. According to Symantec, "the sheer volume of alerts generated by continuous monitoring systems can lead to alert fatigue, making it difficult for security teams to prioritize and respond to the most critical threats."[8]

Another challenge is the cost of implementing continuous monitoring and improvement strategies. Organizations should invest in advanced technologies, such as security information and event management (SIEM) systems, intrusion detection systems (IDS), and behavioral analytics, to support real-time threat detection and response. For smaller organizations with limited resources, the cost of these tools can be a significant barrier. However, as ENISA notes, "the long-term benefits of continuous monitoring, including reduced breach detection times and improved incident response, often outweigh the initial investment."[9]

The Future of Continuous Monitoring and Improvement

As cyber threats continue to evolve, the need for continuous monitoring and improvement will only become more critical. Advances in artificial intelligence (AI) and machine learning (ML) will play a key role in enhancing the capabilities of continuous monitoring systems, enabling them to detect anomalies and predict potential threats with greater accuracy. According to Gartner, "AI and machine learning will enable continuous monitoring systems to learn from past incidents and adapt their defenses in real time, providing organizations with a more proactive approach to cybersecurity."[10]

In the future, organizations will also need to focus on integrating continuous monitoring with broader business continuity and disaster recovery plans. By aligning monitoring efforts with risk management and resilience strategies, organizations can ensure that they are prepared to detect and respond to cyberattacks, capable of recovering quickly, **and minimize** disruption to their operations.

Conclusion

Continuous monitoring and improvement are essential components of an organization's journey toward cyber resilience. By implementing real-time monitoring systems, automating response processes, and continuously refining security practices, organizations can stay ahead of evolving cyber threats and minimize the impact of cyber incidents. As cybercriminals continue to develop new attack techniques, organizations should remain proactive in their approach to cybersecurity, constantly adapting and improving to ensure long-term security and resilience.

11.5 Collaborative Defense and Information Sharing

Collaborative defense and information sharing have become vital to modern cybersecurity strategies. In an era of increasingly complex and interconnected cyber threats, organizations cannot afford to operate in isolation. Cyberattacks often target multiple entities across various sectors simultaneously, exploiting vulnerabilities that span industries and regions. As a result, collaborative defense which is, the practice of working with other organizations to share threat intelligence and coordinate responses, has emerged as a key method for improving overall cyber resilience. According to ENISA, "collaborative defense enables organizations to pool their resources, knowledge, and capabilities, resulting in a more robust and coordinated approach to cybersecurity."[1]

The Importance of Collaborative Defense

Collaborative defense involves sharing cyber threat intelligence, best practices, and strategies across organizations, industries, and even governments. By working together, organizations can gain a more comprehensive understanding of the threat landscape, identify new attack vectors, and share insights about how to mitigate emerging threats. According to Symantec, "no organization, no matter how well resourced, can defend against today's sophisticated cyber threats on its own; collaborative defense is essential for staying ahead of attackers."[2]

Collaboration allows organizations to learn from the experiences of others, particularly those in similar industries or facing common threats. For instance, companies in the financial services sector may face similar threats, such as DDoS attacks or fraud schemes. By sharing insights and defensive strategies, they can develop more effective solutions to protect their networks and customer data. This collective approach to security increases situational awareness and enables a more proactive response to emerging threats.

Benefits of Information Sharing

One key aspect of collaborative defense is information sharing, which involves exchanging data on threats, vulnerabilities, indicators of compromise (IOCs), and attack methods. By sharing information, organizations can improve their threat detection capabilities, shorten the time to incident response, and reduce the overall impact of cyberattacks. According to Palo Alto Networks, "timely sharing of threat intelligence allows organizations to identify attack patterns early and deploy countermeasures more effectively."[3]

Information sharing is particularly important in detecting nation-state attacks and advanced persistent threats (APTs). State-sponsored actors often orchestrate these highly sophisticated, long-term attacks and can target multiple organizations within an industry or geographic region. Collaborative defense enables organizations to exchange insights about the tactics, techniques, and procedures (TTPs) used by these attackers, allowing them to develop more comprehensive defenses. As noted by FireEye, "APT groups often recycle their methods across multiple targets, so sharing information about past attacks can help other organizations better prepare for similar threats."[4]

In addition, real-time information sharing can significantly improve incident response times. For example, if one organization detects an attack in progress, sharing that information with others can enable them to implement preventive measures before the attack spreads. This collaborative approach enhances the ability to contain and mitigate cyber threats on a broader scale.

Cybersecurity Information Sharing Platforms and Initiatives

Several information-sharing platforms and initiatives have been developed to facilitate collaborative defense. One such platform is the Information Sharing and Analysis Center (ISAC), which was established to enable organizations within specific industries to share threat intelligence and collaborate on cybersecurity efforts. ISACs exist in various sectors, including financial services, energy, healthcare, and manufacturing. According to (ISC)², "ISACs provide a trusted environment where organizations can share sensitive information about cyber threats without fear of exposure or legal repercussions."[5]

Another key initiative is the Cybersecurity Information Sharing Act (CISA) 2015, which was enacted in the United States to promote the sharing of cyber threat information between the public and private sectors. The act encourages companies to share information with government agencies such as the Department of Homeland Security (DHS), allowing the government to distribute threat intelligence to other private entities and improve national cybersecurity defenses. According to DHS, "CISA has been instrumental in facilitating the flow of threat intelligence between the public and private sectors, resulting in faster detection and mitigation of cyberattacks."[6]

In Europe, the NIS Directive (Network and Information Security Directive) aims to improve cybersecurity information exchange between EU member states and critical infrastructure operators. The directive mandates that certain organizations, such as energy providers and transportation companies, report cyber incidents to national authorities and share threat intelligence with other operators in their sector.

The Role of Public-Private Partnerships

Public-private partnerships (PPPs)play a crucial role in fostering collaborative defense and information sharing. Governments and private companies should work together to address the complex challenges posed by cyber threats. PPPs enable the sharing of resources, expertise, and threat intelligence to enhance both national and global cybersecurity efforts. According to Deloitte, "effective PPPs allow governments to leverage the innovation and agility of the private sector while providing companies with access to government intelligence and resources."[7]

One notable example of a successful public-private partnership is the National Cyber Security Centre (NCSC) in the United Kingdom, which works with private companies to share threat intelligence and develop coordinated responses to cyber threats. The NCSC's Active Cyber Defence program focuses on providing public and private sector organizations with tools and advice to improve their defenses and share information about cyberattacks. Similarly, the Cybersecurity and Infrastructure Security Agency (CISA) in the United States works closely with private companies to protect critical infrastructure from cyberattacks.

Challenges in Collaborative Defense and Information Sharing

While the benefits of collaborative defense and information sharing are clear, there are challenges. One of the primary challenges is trust. Many organizations are hesitant to share sensitive information about cyber incidents for fear of reputational damage, legal liabilities, or the potential misuse of the information by competitors. According to McAfee, "building trust among participants is essential for successful information sharing, as organizations should feel confident that their data will be protected and used responsibly."[8]

Additionally, legal and regulatory barriers can complicate information sharing, particularly when it involves cross-border exchanges. Privacy regulations such as the GDPR in Europe place strict limits on sharing personal data, which can hinder the exchange of critical information during cyber incidents. Organizations should navigate these regulations carefully to comply with local laws while participating in collaborative defense efforts.

Another challenge is the standardization of shared information. Different organizations may use varying formats, taxonomies, and levels of detail when sharing threat intelligence, making it difficult to analyze and act on the data. To address this, several organizations are working to develop standardized formats for sharing cyber threat information, such as the Structured Threat Information Expression (STIX) and Trusted Automated eXchange of Indicator Information (TAXII) protocols. These standards facilitate the automated exchange of threat intelligence across platforms, improving the efficiency and effectiveness of information sharing.

Future Trends in Collaborative Defense

Looking ahead, the future of collaborative defense will likely involve greater automation and the use of artificial intelligence (AI) to analyze shared threat intelligence more quickly and accurately. AI-driven platforms can automatically process large volumes of shared data, identify patterns, and recommend responses to cyber threats. According to Gartner, "AI will play a critical role in enhancing the effectiveness of collaborative defense by enabling faster analysis of shared threat intelligence and providing actionable insights in real time."[9]

Furthermore, as cyber threats continue to evolve and target critical infrastructure sectors, cross-sector collaboration will be emphasized. Organizations in different industries will need to share information and collaborate on defense strategies to protect the interconnected systems that power financial services, energy grids, healthcare, and transportation networks.

Conclusion

Collaborative defense and information sharing are essential for building a robust and resilient cybersecurity ecosystem. Organizations can pool their knowledge, resources, and capabilities by working together to detect, mitigate, and prevent cyberattacks more effectively. Although challenges such as trust and regulatory barriers exist, the benefits of collaborative defense far outweigh these obstacles

Test Your knowledge on Building Cyber Resilience

Q1: What does it mean to integrate cybersecurity into an organization's DNA?

Q2: How does adaptive security architecture enhance cyber resilience?

Q3: Why is continuous monitoring important for maintaining cybersecurity?

Q4: What role does continuous improvement play in cybersecurity?

Q5: How does collaborative defense contribute to cybersecurity resilience?

Q6: What are the benefits of information sharing in cybersecurity?

Q7: How do public-private partnerships support cybersecurity efforts?

Q8: What challenges do organizations face in implementing continuous monitoring systems?

Q9: How does continuous risk assessment improve cyber resilience?

Q10: How can organizations ensure the success of their adaptive security strategies?

Q11: What role does AI play in real-time threat detection?

Q12: How can organizations overcome the trust barriers in information sharing?

Q13: What are the long-term benefits of continuous improvement in cybersecurity?

Q14: How do cybersecurity maturity models contribute to continuous improvement?

Q15: How can organizations measure the effectiveness of their cyber resilience efforts?

ANSWERS

Q1: A: Integrating cybersecurity into an organization's DNA means embedding security practices into every aspect of the organization, from leadership to daily operations, ensuring that security is a core value.

Q2: A: Adaptive security architecture allows organizations to continuously assess and adjust their defenses based on evolving threats, providing a more dynamic and responsive approach to security.

Q3: A: Continuous monitoring enables organizations to detect and respond to threats in real time, reducing the time attackers have to cause damage and improving overall resilience.

Q4: A: Continuous improvement involves regularly updating security policies, tools, and processes based on new threats and vulnerabilities, ensuring that an organization's defenses remain robust over time.

Q5: A: Collaborative defense enables organizations to share threat intelligence and resources, helping them respond more effectively to cyber threats by pooling their knowledge and capabilities.

Q6: A: Information sharing improves threat detection, speeds up response times, and helps organizations prevent attacks by learning from the experiences of others.

Q7: A: Public-private partnerships allow governments and private organizations to share intelligence, coordinate responses, and enhance national cybersecurity through collaborative defense strategies.

Q8: A: Challenges include processing the large volumes of data generated by monitoring tools, managing alert fatigue, and the cost of implementing advanced monitoring technologies.

Q9: A: Continuous risk assessment helps organizations identify emerging vulnerabilities and prioritize defenses based on the changing threat landscape, ensuring that security efforts are focused on the most pressing risks.

Q10: A: Success requires integrating **AI**, machine learning, and threat intelligence into adaptive security systems while continuously updating defense mechanisms to keep up with new threats.

Q11: A: AI enables real-time analysis of network traffic and system behavior, identifying threats as they occur and allowing organizations to respond quickly to minimize damage.

Q12: A: Building trust in information sharing requires clear guidelines on data protection, legal assurances against misuse, and fostering a culture of collaboration among participants.

Q13: A: Continuous improvement ensures that organizations can adapt to new threats, maintain compliance with evolving regulations, and build a stronger, more resilient security posture over time.

Q14: A: Maturity models help organizations benchmark their security practices, identify gaps, and create a roadmap for improving cybersecurity capabilities over time.

Q15: A: Effectiveness can be measured through metrics like incident response time, mean time to recovery (MTTR), the number of successful attacks, and the ability to maintain operations during a cyber incident.

A-Z ACRONYMS AND ABBREVIATIONS

1. **AI** – Artificial Intelligence
2. The simulation of human intelligence processes by machines, especially computer systems.
3. **API** – Application Programming Interface
4. A set of protocols and tools for building software and enabling communication between different applications.
5. **APT** – Advanced Persistent Threat
6. A long-term, targeted cyberattack where an unauthorized user gains continuous access to a network.
7. **APTs** – Advanced Persistent Threats
8. Multiple or prolonged targeted cyberattacks that remain undetected over an extended period.
9. **BIA** – Business Impact Analysis
10. A process to identify and evaluate the effects of disruptions to business operations.
11. **BYOD** – Bring Your Own Device
12. A policy allowing employees to use personal devices for work-related tasks, often presenting security risks.
13. **CCTV** – Closed-Circuit Television
14. A video surveillance system used for monitoring, often integrated with security protocols.
15. **CIRT** – Cyber Incident Response Team
16. A team responsible for responding to cybersecurity incidents and mitigating their effects.
17. **CISO** – Chief Information Security Officer
18. A senior executive responsible for overseeing an organization's information security strategy and operations.
19. **CSIRT** – Computer Security Incident Response Team

20. A group of experts that responds to and manages cybersecurity incidents for an organization or government body.
21. **CVE** – Common Vulnerabilities and Exposures
22. A system and database used to categorize and identify known security vulnerabilities.
23. **DDoS** – Distributed Denial of Service
24. An attack that makes a system or network unavailable by overwhelming it with traffic from multiple sources.
25. **DMZ** – Demilitarized Zone
26. A subnetwork designed to separate internal systems from untrusted networks like the public internet.
27. **DoS** – Denial of Service
28. An attack aimed at making a network or system unavailable by flooding it with requests.
29. **E2EE** – End-to-End Encryption
30. A communication method where only the communicating users can read the messages, ensuring privacy.
31. **FBI** – Federal Bureau of Investigation
32. The U.S. domestic intelligence and security service, also responsible for investigating cybercrime.
33. **FISMA** – Federal Information Security Management Act
34. U.S. legislation that mandates protection of federal information systems.
35. **GDPR** – General Data Protection Regulation
36. An EU law on data protection and privacy that applies to all individuals within the European Union.
37. **HIPAA** – Health Insurance Portability and Accountability Act
38. U.S. law that sets standards for protecting sensitive patient information.
39. **ICS** – Industrial Control System
40. A system used to control industrial processes, often targeted by cyberattacks due to its critical nature.
41. **IDS** – Intrusion Detection System
42. A system that monitors network traffic for suspicious activity and alerts the system administrators.
43. **IoT** – Internet of Things
44. A network of interconnected devices that can collect and exchange data via the internet.
45. **IPS** – Intrusion Prevention System
46. A system that monitors and actively prevents malicious activities in real-time.
47. **IR** – Incident Response
48. The process of identifying, managing, and recovering from a cybersecurity breach or attack.
49. **ISO** – International Organization for Standardization
50. An international body that develops and publishes global standards, including for cybersecurity.

51. **ISP** – Internet Service Provider
52. A company that provides individuals and businesses with internet access, often involved in managing network security.
53. **MFA** – Multi-Factor Authentication
54. A security method that requires more than one form of identification to verify a user's identity.
55. **MOU** – Memorandum of Understanding
56. A document outlining an agreement between parties, often used in cybersecurity collaborations.
57. **MSSP** – Managed Security Service Provider
58. A company that provides outsourced monitoring and management of security systems for businesses.
59. **MTTD** – Mean Time to Detect
60. The average time it takes to detect a cyberattack or breach.
61. **MTTR** – Mean Time to Recovery
62. The average time required to recover from a system failure or cyberattack.
63. **NGFW** – Next-Generation Firewall
64. An advanced type of firewall that offers features like deep packet inspection and application-level filtering.
65. **NIST** – National Institute of Standards and Technology
66. A U.S. federal agency that develops cybersecurity frameworks and standards.
67. **PAM** – Privileged Access Management
68. A system that controls and monitors access to sensitive systems or data by authorized users.
69. **PCI DSS** – Payment Card Industry Data Security Standard
70. A set of security standards to protect credit card data during transactions.
71. **PEN Testing** – Penetration Testing
72. A method of testing a system or network's security by simulating an attack to identify vulnerabilities.
73. **RAT** – Remote Access Trojan
74. A type of malware that allows attackers to gain unauthorized remote control over a system.
75. **RBAC** – Role-Based Access Control
76. A method of limiting access to systems based on a user's role within an organization.
77. **ROI** – Return on Investment
78. A performance metric used to evaluate the efficiency or profitability of a cybersecurity investment.
79. **RPO** – Recovery Point Objective
80. The maximum acceptable amount of data loss, measured in time, that an organization is willing to tolerate after a failure.
81. **RTO** – Recovery Time Objective
82. The maximum amount of time that a system can be down before it severely impacts business operations.

83. **SLA** – Service Level Agreement
84. A contract that defines the level of service expected between a provider and a client, including security standards.
85. **SIEM** – Security Information and Event Management
86. A technology platform that provides real-time analysis of security alerts generated by network hardware and applications.
87. **SOC** – Security Operations Center
88. A centralized unit within an organization responsible for monitoring, detecting, and responding to cybersecurity incidents.
89. **TFA** – Two-Factor Authentication
90. A security method requiring two forms of verification before granting access to a system.
91. **TTPs** – Tactics, Techniques, and Procedures
92. The behaviors and methods associated with a particular cyberattack or threat actor.
93. **VPN** – Virtual Private Network
94. A service that encrypts a user's internet connection to protect their data and privacy.
95. **WAF** – Web Application Firewall
96. A firewall specifically designed to protect web applications by filtering and monitoring HTTP traffic.

GLOSSARY

Below is a comprehensive glossary, broken down by chapter, covering the key terms, concepts, and technologies introduced throughout the book. Each definition is tailored to the topics explored in that chapter.

Chapter 1: Foundations of Cybersecurity

1. **Confidentiality**: The principle of ensuring that sensitive data is accessed only by authorized users and is protected from unauthorized disclosure.
2. **Integrity**: Ensuring the accuracy and trustworthiness of data by preventing unauthorized alterations.
3. **Availability**: The guarantee that authorized users have timely and reliable access to data and systems when needed.
4. **Phishing**: A type of social engineering attack where an attacker masquerades as a trustworthy entity to trick individuals into divulging sensitive information.
5. **Malware**: Malicious software designed to disrupt, damage, or gain unauthorized access to a computer system.
6. **Social Engineering**: Psychological manipulation of individuals into performing actions or divulging confidential information for malicious purposes.
7. **Zero-Day Exploit**: A cyberattack that occurs on the same day a vulnerability is discovered and before a patch or fix can be applied.
8. **Encryption**: The process of converting data into a coded form to prevent unauthorized access.
9. **Multi-Factor Authentication (MFA)**: A security system that requires more than one method of authentication from independent categories of credentials to verify the user's identity.
10. **Firewall**: A network security system that monitors and controls incoming and outgoing traffic based on predetermined security rules.

Chapter 2: Technological Innovations in Cybersecurity

1. **Artificial Intelligence (AI)**: The simulation of human intelligence in machines programmed to think and learn, increasingly used in cybersecurity for threat detection.
2. **Machine Learning (ML)**: A subset of AI that allows systems to learn from data, identify patterns, and make decisions with minimal human intervention.
3. **Blockchain**: A decentralized, distributed digital ledger that records transactions in a secure, transparent, and immutable way. It is used in cybersecurity to secure transactions and data integrity.
4. **Quantum Computing**: A type of computing that uses quantum-mechanical phenomena, such as superposition and entanglement, to perform operations, posing both opportunities and threats to encryption.
5. **End-to-End Encryption (E2EE)**: A method of securing communication where only the communicating users can read the messages, ensuring privacy from third parties.
6. **Cloud Security**: The technologies and policies designed to protect data and applications stored in cloud environments from cyber threats.
7. **5G Networks**: The fifth generation of mobile network technology, which promises faster speeds and lower latency but introduces new security challenges due to its increased connectivity.
8. **Edge Computing**: A distributed computing paradigm that brings computation and data storage closer to the location where it is needed, increasing both performance and security challenges.
9. **Biometric Authentication**: A security process that uses biological characteristics, such as fingerprints or facial recognition, to authenticate individuals.
10. **Quantum Cryptography**: A form of encryption that uses the principles of quantum mechanics to secure communication, expected to provide unbreakable encryption.

Chapter 3: The Cyber Threat Landscape

1. **Advanced Persistent Threat (APT)**: A prolonged and targeted cyberattack in which an intruder gains access to a network and remains undetected for an extended period.

2. **Ransomware**: A type of malware that encrypts the victim's files, demanding a ransom payment to restore access.
3. **Insider Threat**: A threat posed by individuals within an organization who exploit their authorized access to harm the organization's data, systems, or operations.
4. **Distributed Denial of Service (DDoS)**: An attack that overwhelms a system with excessive traffic, rendering it unavailable to legitimate users.
5. **Trojans**: Malware that disguises itself as legitimate software but, once executed, gives attackers access to the system.
6. **Worms**: Self-replicating malware that spreads across networks without needing a host file or human interaction.
7. **Botnet**: A network of compromised devices controlled remotely by a hacker to launch large-scale attacks, such as DDoS attacks.
8. **Phishing**: The fraudulent attempt to obtain sensitive information by pretending to be a trustworthy entity in electronic communications.
9. **Spear Phishing**: A targeted phishing attack aimed at a specific individual or organization, often personalized with specific details about the victim.
10. **IoT (Internet of Things)**: A system of interrelated computing devices, machines, and objects with unique identifiers and the ability to transfer data over a network, often vulnerable to attacks.

Chapter 4: Cybersecurity for Individuals

1. **Personal Data**: Information that can be used to identify an individual, such as names, addresses, phone numbers, and social security numbers.
2. **Password Manager**: A software application that helps users generate, store, and manage secure passwords for their online accounts.
3. **Digital Hygiene**: The practice of maintaining good security habits, such as updating software, using strong passwords, and avoiding suspicious links to reduce cybersecurity risks.
4. **Home Network Security**: Practices and technologies designed to protect a home's internet-connected devices, such as changing router passwords and enabling encryption.
5. **Virtual Private Network (VPN)**: A service that encrypts internet connections and masks a user's IP address to protect their data from surveillance and tracking.
6. **Multi-Factor Authentication (MFA)**: A security system requiring two or more authentication factors, such as a password and a fingerprint, to verify identity.

7. **Phishing Scam**: A fraudulent attempt to obtain sensitive information through deceptive emails, messages, or websites.
8. **Encryption**: The process of encoding information so that only authorized parties can read it.
9. **Data Breach**: An incident where unauthorized individuals gain access to sensitive, confidential, or protected data.
10. **Two-Factor Authentication (2FA)**: A subset of MFA that requires two distinct forms of identification to access an account, such as a password and a one-time code.

Chapter 5: Organizational Cybersecurity

1. **Risk Assessment**: The process of identifying, evaluating, and prioritizing risks to minimize, monitor, and control the impact of security threats.
2. **Incident Response Plan**: A set of procedures to detect, respond to, and recover from security breaches, ensuring minimal damage and disruption.
3. **Business Continuity**: The capability of an organization to continue delivery of services and products during or after a disruptive event, including cyber incidents.
4. **Security Policy**: A document that outlines how an organization protects its information and technology assets, including roles and responsibilities for maintaining security.
5. **Privileged Access Management (PAM)**: A system for controlling and monitoring access to critical systems and sensitive data, often used to mitigate insider threats.
6. **Third-Party Risk Management**: The practice of identifying and mitigating risks posed by external vendors, contractors, or partners who have access to an organization's network or data.
7. **Zero Trust Architecture**: A security model that assumes no network or user is trusted by default, requiring verification for every attempt to access resources.
8. **Security Awareness Training**: Training provided to employees to make them aware of potential cybersecurity threats and how to avoid them.
9. **Security Governance**: The framework that ensures an organization's security policies align with its business objectives and regulatory requirements.
10. **Patch Management**: The process of managing software updates to correct vulnerabilities and improve security.

Chapter 6: Cybersecurity in Key Industries

1. **PCI DSS (Payment Card Industry Data Security Standard)**: A set of security standards designed to ensure that all companies accepting, processing, storing, or transmitting credit card information maintain a secure environment.
2. **HIPAA (Health Insurance Portability and Accountability Act)**: U.S. legislation providing data privacy and security provisions for safeguarding medical information.
3. **Industrial Control Systems (ICS)**: Systems used in industrial sectors to control processes, often targeted in cyberattacks aiming to disrupt critical infrastructure.
4. **Electronic Health Records (EHR)**: Digital records of patients' medical histories, diagnoses, treatments, and other healthcare information, requiring strong security measures under HIPAA.
5. **Smart Factory**: A highly digitized and connected production facility that relies on smart manufacturing technologies to improve efficiency, requiring advanced security to protect its operations.
6. **Data Privacy**: The practice of protecting personal and sensitive data from unauthorized access or disclosure, which is particularly important in healthcare and financial services.
7. **Critical Infrastructure**: Systems and assets essential to national security and public safety, such as energy grids, financial institutions, and transportation systems, which are often targeted by cyberattacks.
8. **Tokenization**: A process that replaces sensitive data with unique identification symbols, or tokens, that retain the essential information but are less vulnerable to exposure.
9. **Cloud-Based Security**: Security services and solutions that protect cloud-based systems and data from breaches, loss, or other threats.
10. **Role-Based Access Control (RBAC)**: A system that restricts system access to authorized users based on their role within the organization, ensuring only those with a need to know can access sensitive data.

Chapter 7: The Economics of Cybersecurity

1. **Return on Investment (ROI)**: A measure used to evaluate the efficiency of an investment, particularly in cybersecurity, by comparing the gains (e.g., cost savings from avoided breaches) to the investment costs.

2. **Cyber Insurance**: Insurance products are designed to protect businesses from losses related to cyberattacks, such as data breaches, ransomware, and operational disruptions.

3. **Direct Costs**: The immediate costs of a cyberattack, including breach detection, incident response, legal fees, and paying ransoms.

4. **Indirect Costs**: The longer-term impacts of a cyberattack, such as loss of customer trust, brand damage, and loss of revenue from downtime.

5. **Ransomware Payments**: Payments made to cybercriminals to regain access to encrypted data or systems, though paying does not guarantee recovery and is discouraged by law enforcement.

6. **Risk Management**: The process of identifying, assessing, and prioritizing cybersecurity risks and applying resources to minimize or mitigate them.

7. **Cybersecurity Investment Strategy**: A plan that allocates financial and human resources to different cybersecurity tools, technologies, and practices to maximize protection while minimizing costs.

8. **Cybercrime Economy**: The illicit market for stolen data, hacking tools, malware, and other cybercrime-related services, often facilitated through the dark web.

9. **Cybersecurity Spending**: The allocation of financial resources to protect an organization's data, networks, and systems from cyberattacks, including investments in software, hardware, and human resources.

10. **Risk Transfer**: A strategy where organizations purchase insurance or outsource security operations to mitigate the financial impact of a cyberattack.

11. **Business Impact Analysis (BIA)**: A process that assesses the potential effects of an interruption to critical business operations as a result of a cyber incident.

12. **Cybersecurity Skills Gap**: The shortage of qualified professionals in the cybersecurity field affects an organization's ability to defend against threats.

13. **Security Automation**: The use of automated tools to detect, respond to, and mitigate cyber threats without the need for constant human intervention.

14. **Dark Web**: A part of the internet that requires special software (e.g., Tor) to access, often used by cybercriminals to conduct illegal activities such as selling stolen data or hacking tools.

15. **Economic Incentives**: Financial motivations provided by governments or industries, such as tax breaks or grants, to encourage organizations to improve their cybersecurity measures.

Chapter 8: Ethical and Legal Aspects of Cybersecurity

1. **Privacy vs. Security**: The balance between protecting individual privacy rights and ensuring security, particularly in the context of surveillance and data protection.
2. **Ethical Hacking**: The practice of testing a system's defenses by simulating cyberattacks with the permission of the system owner, also known as **penetration testing**.
3. **Responsible Disclosure**: The ethical practice of privately informing a company or software vendor of a security vulnerability so it can be fixed before being publicly disclosed.
4. **Government Surveillance**: The monitoring of digital communications and activities by government agencies to protect national security, often controversial due to concerns about civil liberties.
5. **Civil Liberties**: Fundamental individual rights, such as privacy and freedom of expression, which can be impacted by surveillance and other cybersecurity measures.
6. **GDPR (General Data Protection Regulation)**: A European Union law that strengthens data protection rights for individuals and imposes strict obligations on organizations to protect personal data.
7. **HIPAA (Health Insurance Portability and Accountability Act)**: U.S. legislation that sets standards for protecting sensitive patient information and ensuring it remains confidential.
8. **Corporate Responsibility**: The ethical obligation of companies to protect customer data, disclose breaches, and be transparent about security incidents.
9. **Cybersecurity Laws**: Regulations at the national and international level that govern the responsibilities of organizations in protecting data and reporting breaches.
10. **International Cybersecurity Agreements**: Treaties and agreements between nations to cooperate on issues related to cybersecurity, cybercrime, and the protection of critical infrastructure.
11. **Cyber Warfare**: The use of cyberattacks by a nation-state to disrupt or damage another nation's infrastructure, including energy grids, financial systems, and military networks.

12. **AI Ethics**: The ethical considerations around the use of AI in cybersecurity, such as ensuring fairness, transparency, and avoiding bias in automated decision-making.
13. **Data Breach Notification Laws**: Regulations that require organizations to notify individuals and regulators when personal data is compromised, often with a defined timeframe for disclosure.
14. **Cyber Diplomacy**: The practice of using diplomatic channels to address and resolve international cybersecurity issues, such as cross-border cyberattacks and cybercrime.
15. **Cybersecurity Liability**: The legal responsibility organizations face for failing to protect customer data or prevent cyberattacks, often leading to lawsuits or regulatory fines.

Chapter 9: Global Cybersecurity Landscape

1. **Cyber Warfare**: The use of digital attacks by nation-states to destroy, damage, or disrupt the information systems of other countries, typically as part of a larger conflict.
2. **Nation-State Actor**: A government or its agents engaging in cyber espionage, sabotage, or cyber warfare, often targeting other governments or critical infrastructure.
3. **International Cooperation**: The collaboration between countries to combat cybercrime, share intelligence, and establish frameworks for addressing global cyber threats.
4. **Cybersecurity Treaties**: Formal agreements between nations that outline protocols for dealing with cyber incidents, cyber defense strategies, and cybercrime enforcement.
5. **Comparative Analysis of National Cybersecurity Strategies**: A method of evaluating how different countries approach cybersecurity, comparing their policies, investments, and regulatory frameworks.
6. **Cyber Diplomacy**: Diplomatic efforts focused on establishing international norms for behavior in cyberspace and fostering cooperation to prevent cyber conflicts.
7. **Cross-Border Data Flow**: The movement of data across national borders raises legal and security challenges, particularly in terms of data sovereignty and privacy regulations.
8. **Jurisdictional Issues**: Legal challenges that arise when cybercrimes cross international borders, making it difficult to determine which country has the authority to prosecute.

9. **Cybersecurity Culture**: The collective mindset, values, and behaviors related to security within an organization or country contribute to how seriously cybersecurity is taken.

10. **Data Sovereignty**: The concept that data is subject to the laws and governance structures of the country in which it is collected or processed, which can complicate cross-border data transfers.

11. **Cyber Espionage**: The practice of using cyber means to spy on other countries or organizations to gather intelligence, often for economic or political advantage.

12. **Digital Sovereignty**: The ability of a nation to control its own digital infrastructure, data, and cyberspace, free from external influence or cyberattacks.

13. **Global Cybersecurity Standards**: Internationally recognized guidelines and frameworks that organizations and countries follow to secure their digital assets and protect against cyber threats.

14. **Mutual Legal Assistance Treaties (MLATs)**: Agreements between countries to cooperate in investigating and prosecuting cross-border crimes, including cybercrime.

15. **Cybersecurity Norms**: Internationally agreed-upon principles and rules that guide responsible behavior in cyberspace, aimed at reducing the risk of cyber conflict.

Chapter 10: The Future of Cybersecurity

1. **Quantum Cryptography**: A method of securing data using the principles of quantum mechanics, offering theoretically unbreakable encryption that is immune to current hacking techniques.

2. **AI-Powered Attacks**: Cyberattacks that leverage artificial intelligence to identify vulnerabilities, automate attacks, or evade detection by adaptive security systems.

3. **Predictive Cybersecurity**: A proactive approach that uses data analytics and machine learning to predict future cyber threats and vulnerabilities, allowing organizations to defend against them in advance.

4. **5G Security**: The challenges and opportunities associated with securing the fifth generation of mobile network technology, which connects billions of devices globally.

5. **Edge Computing**: A model of computing where data is processed at the edge of a network, closer to the source of the data, improving response times but also introducing new security risks.

6. **Proactive Threat Hunting:** The practice of actively searching for cyber threats within an organization's network before they manifest as breaches or incidents.

7. **Autonomous Systems:** Systems, such as self-driving cars or drones, that operate independently using AI and machine learning pose unique cybersecurity challenges due to their reliance on interconnected systems.

8. **Post-Quantum Cryptography:** Encryption methods that are designed to be secure against the potential capabilities of quantum computing, offering protection in a future where traditional cryptographic methods may be vulnerable.

9. **AI-Driven Defenses:** The use of artificial intelligence to enhance cybersecurity defenses, enabling faster threat detection, real-time response, and automated mitigation of attacks.

10. **Cybersecurity Innovation:** The development of new technologies, strategies, and methodologies aimed at improving the ability to defend against evolving cyber threats.

11. **Zero Trust Security:** A cybersecurity framework that requires continuous verification of all users, devices, and systems, regardless of whether they are inside or outside the network perimeter.

12. **Autonomous Cyber Defense:** The use of AI and machine learning to create self-sustaining defense mechanisms that can automatically detect, respond to, and neutralize threats without human intervention. And **Autonomous Cyberattack** is the opposite.

13. **Quantum-Resistant Encryption:** Encryption algorithms that are designed to withstand attacks from quantum computers, ensuring long-term data security even as quantum technology advances.

14. **AI Ethics in Cybersecurity:** The ethical concerns surrounding the use of AI in cybersecurity, including issues related to privacy, bias, and the potential for AI to be used maliciously.

15. **Cybersecurity Forecasting:** The practice of predicting future cyber threats based on current trends and data analytics, allowing organizations to take preventive measures before an attack occurs.

Chapter 11: Building Cyber Resilience

1. **Cyber Resilience:** The ability of an organization to prepare for, respond to, and recover from cyberattacks while maintaining critical business functions.

2. **Adaptive Security Architecture**: A security model that continuously evolves and adapts to new threats, allowing for real-time detection, response, and mitigation.
3. **Continuous Monitoring**: The practice of constantly monitoring an organization's network, systems, and data for signs of unauthorized access, anomalies, or potential breaches.
4. **Continuous Improvement**: An approach to cybersecurity that emphasizes regular updates, reviews, and enhancements to security measures in response to emerging threats.
5. **Collaborative Defense**: The practice of sharing threat intelligence and resources among organizations, industries, and governments to improve overall cybersecurity defense.
6. **Threat Intelligence Sharing**: The exchange of information about cyber threats, vulnerabilities, and attack methods between organizations to strengthen their defenses.
7. **Public-Private Partnership (PPP)**: Collaboration between government agencies and private companies to improve cybersecurity, share threat intelligence, and respond to cyber incidents.
8. **Security Information and Event Management (SIEM)**: A technology that provides real-time analysis of security alerts generated by applications and network hardware, often used in continuous monitoring.
9. **Incident Response Automation**: The use of automated systems to detect and respond to cyber incidents, minimizing the time required for human intervention.
10. **Maturity Models**: Frameworks that assess the maturity of an organization's cybersecurity practices and provide a roadmap for continuous improvement.
11. **Behavioral Analytics**: A cybersecurity technique that analyzes the behavior of users and systems to detect anomalies that may indicate a security threat.
12. **Forensic Analysis**: The investigation of security incidents to understand how they occurred, who was responsible, and what vulnerabilities were exploited.
13. **Security Awareness**: The collective understanding within an organization about the importance of cybersecurity and the role each employee plays in maintaining it.

14. **Risk-Based Security**: A security strategy that prioritizes the protection of assets based on the level of risk they pose to the organization, focusing resources where they are most needed.

15. **Red Teaming**: A security exercise where ethical hackers attempt to breach an organization's defenses to identify weaknesses and improve the overall security posture.

REFERENCES AND

CITATIONS

FOOTNOTES

CHAPTER 1
FOUNDATIONS OF CYBERSECURITY

1.1 The Evolution of Cybersecurity
References Footnotes

1. Peter Neumann, quoted in "The Morris Worm: A Historic Lesson in Cybersecurity," *Cybersecurity Journal*, vol. 32, no. 1 (2020): 18-25.
2. Symantec, "The History of Norton Antivirus," *Symantec Security Whitepaper*, accessed October 6, 2024, https://www.symantec.com/norton/history.
3. Bruce Schneier, *Applied Cryptography* (New York: Wiley, 1996).
4. "Target Breach Highlights Growing Threat of Cyber Attacks," *New York Times*, January 14, 2014.
5. Scott Shackelford, *Managing Cyber Attacks in International Law, Business, and Relations: In Search of Cyber Peace* (New York: Cambridge University Press, 2014), 54.
6. "The Future of Quantum Computing and Cybersecurity," *Journal of Advanced Computing*, vol. 45, no. 2 (2022): 102-120.

1.2 Key Concepts and Terminology
References Footnotes

1. "Target Breach Highlights Growing Threat of Cyber Attacks," *New York Times*, January 14, 2014.
2. Matthew Bishop, *Computer Security: Art and Science* (Addison-Wesley, 2018), 34.
3. "Bangladesh Bank Cyber Heist: A $101 Million Mystery," *BBC News*, March 8, 2016.
4. "The Dyn DNS Attack: The Day the Internet Died," *Wired*, October 26, 2016.
5. "How MFA Prevents Phishing Attacks: A Case Study," *Cybersecurity Journal*, vol. 42, no. 3 (2020): 56-67.
6. Edward Snowden, *Permanent Record* (Metropolitan Books, 2019).
7. Bruce Schneier, *Applied Cryptography* (Wiley, 1996).
8. "2023 Encryption Trends," *Ponemon Institute Survey*, accessed October 6, 2024.
9. "Next-Generation Firewalls: Bridging the Gap," *Network World*, March 14, 2020.
10. Martin Roesch, "Snort: IDS for the Masses," *SecurityFocus*, November 22, 2004.
11. "The Rise of Zero Trust Security: Forrester Report," *Forrester*, 2023.
12. "BeyondCorp: A New Approach to Enterprise Security," *Google Cloud Whitepaper*, accessed October 6, 2024.

1.3 The CIA Triad: Confidentiality, Integrity, and Availability
Footnotes

1. Eric Cole, *Advanced Persistent Threat: Understanding the Danger and How to Protect Your Organization* (Syngress, 2012).
2. "Yahoo Data Breach: Largest Hack in History," *The Guardian*, December 14, 2016.
3. Bruce Schneier, *Data and Goliath: The Hidden Battles to Collect Your Data and Control Your World* (Norton, 2015).
4. "2021 Global Information Security Workforce Study," *(ISC)²*, accessed October 7, 2024.
5. William Stallings, *Cryptography and Network Security* (Pearson, 2020).
6. "The Bangladesh Bank Cyber Heist: A $81 Million Mystery," *Reuters*, April 6, 2016.
7. "Electronic Health Records: Maintaining Integrity," *Health IT Journal*, vol. 25, no. 2 (2021): 101-115.
8. "The 2016 Dyn DNS Attack: A Major Internet Disruption," *Wired*, October 24, 2016.
9. Marcus Ranum, *The Myth of Homeland Security* (Wiley, 2003).
10. "AWS Outage: The Risks of Cloud Dependence," *TechCrunch*, November 25, 2020.
11. "2021 Data Breach Investigations Report," *Verizon*, accessed October 7, 2024.

1.4 Types of Cyber Threats and Attacks
References Footnotes

1. "WannaCry Ransomware Attack: An International Crisis," *BBC News*, May 12, 2017.
2. H. Saini, Y.S. Rao, and T.C. Panda, "Cyber-crimes and their impacts: A review," *International Journal of Engineering and Technology*, 2012.
3. "Zeus Trojan: A Banking Heist," *Forbes*, September 24, 2013.
4. "Colonial Pipeline Ransomware Attack," *The Washington Post*, May 10, 2021.
5. "DNC Spear-Phishing Attack," *The New York Times*, July 13, 2016.
6. M. Uma and G. Padmavathi, "A survey on various cyber-attacks and their classification," *International Journal of Network Security*, 2013.
7. "Business Email Compromise (BEC) Scams," *FBI Internet Crime Report*, 2020.
8. "Dyn DNS Attack Disrupts Major Websites," *Wired*, October 24, 2016.
9. Ö. Aslan et al., "A comprehensive review of cyber security vulnerabilities, threats, attacks, and solutions," *Electronics*, 2023.
10. "Stuxnet: A New Age of Cyber Warfare," *The Guardian*, September 21, 2010.
11. I. Agrafiotis et al., "A taxonomy of cyber-harms: Defining the impacts of cyber-attacks," *Journal of Cybersecurity*, 2018.
12. Edward Snowden, *Permanent Record*, (New York: Metropolitan Books, 2019).
13. L. Liu et

1.5 Basic Principles of Digital Protection
References Footnotes

1. M. Jouini, L. Rabai, and A. Aissa, "Classification of security threats in information systems," *Procedia Computer Science*, 2014.
2. Gunduz, M. Z., and Das, R., "Analysis of cyber-attacks on smart grid applications," *IEEE International Conference on Artificial Intelligence*, 2018.
3. Uma, M., and Padmavathi, G., "A survey on various cyber-attacks and their classification," *International Journal of Network Security*, 2013.

1.6 The Cybersecurity Ecosystem: Stakeholders and Their Roles
References Footnotes

1. Ponemon Institute, "Cost of a Data Breach Report 2021," IBM Security, 2021.
2. "2021 Data Breach Investigations Report," Verizon, accessed October 7, 2024.
3. Richard A. Clarke, *Cyber War: The Next Threat to National Security and What to Do About It* (HarperCollins, 2010).
4. "Equifax Settles Data Breach Case for $700 Million," *The New York Times*, July 22, 2019.
5. "Cybersecurity and Infrastructure Security Agency (CISA)," U.S. Department of Homeland Security, accessed October 7, 2024.
6. "The SolarWinds Hack: A Comprehensive Timeline," *Wired*, January 10, 2021.
7. "Global Cybersecurity Market Forecast to Reach $400 Billion by 2026," *Forbes*, June 10, 2022.
8. "Unit 42 Threat Intelligence," Palo Alto Networks, accessed October 7, 2024.
9. "Operation DisrupTor: Dark Web Marketplace Takedown," *Europol Press Release*, September 22, 2020.
10. "ITU Cybersecurity Initiatives," International Telecommunication Union, accessed October 7, 2024.
11. "Global Forum on Cyber Expertise (GFCE)," accessed October 7, 2024.

CHAPTER 2
TECHNOLOGICAL INNOVATIONS IN CYBERSECURITY

2.1 Artificial Intelligence and Machine Learning in Threat Detection
References Footnotes

1. Mohammadi, Amin, et al., "AI in Cybersecurity: Threat Detection Systems," *Journal of Network Security*, 2020.
2. "Darktrace: AI-Driven Cybersecurity Solutions," *Darktrace Whitepaper*, accessed October 8, 2024.
3. Chio, Christopher, and David Freeman, *Machine Learning and Security* (O'Reilly Media, 2018).
4. "Detecting Fileless Malware with Neural Networks," *Journal of Malware Analysis*, 2021.
5. "Gmail Blocks Over 100 Million Phishing Emails Every Day," *Google Security Blog*, April 12, 2020.
6. "IBM QRadar SIEM: Real-Time Threat Detection," *IBM Security Whitepaper*, accessed October 8, 2024.
7. Alazab, Mamoun, et al., "Real-Time Cybersecurity: AI for Threat Detection," *IEEE Access*, 2022.
8. West, Darrell M., and Nishith Bhattacharya, "Adversarial AI in Cybersecurity: Challenges and Solutions," *Brookings Institution Report*, 2021.

2.2 Blockchain Technology and Its Security Applications
References Footnotes

1. William Mougayar, *The Business Blockchain: Promise, Practice, and the Application of the Next Internet Technology* (Wiley, 2016).
2. Michael Crosby et al., "Blockchain technology: Beyond Bitcoin," *Applied Innovation Review*, 2016.
3. "Blockchain for Electronic Health Records," *Health IT Journal*, 2020.
4. "Blockchain and DDoS Prevention: A New Approach," *Journal of Network Security*, 2021.
5. Primavera De Filippi and Aaron Wright, *Blockchain and the Law: The Rule of Code* (Harvard University Press, 2018).
6. "Civic: Decentralized Identity Management," Civic Whitepaper, 2021.
7. "Microsoft Azure Blockchain for Decentralized Identity," *Microsoft Blockchain Blog*, 2020.
8. "IBM Food Trust: Blockchain for the Food Industry," IBM Whitepaper, accessed October 8, 2024.
9. Don Tapscott and Alex Tapscott, *Blockchain Revolution: How the Technology Behind Bitcoin Is Changing Money, Business, and the World* (Penguin, 2016).
10. "Ethereum: Smart Contracts and Decentralized Applications," *Ethereum Whitepaper*, accessed October 8, 2024.
11. Vitalik Buterin, "A Next-Generation Smart Contract and Decentralized Application Platform," Ethereum Whitepaper, 2014.
12. "51% Attacks: Risks and Realities," *Blockchain Security Journal*, 2019.

2.3 Cloud Security Advancements
References Footnotes

1. Gartner, "Cloud Security Predictions 2025," *Gartner Report*, 2021.
2. "Capital One Data Breach: Lessons Learned," *The New York Times*, August 1, 2019.
3. "Amazon Web Services: Encryption in the Cloud," *AWS Security Whitepaper*, 2020.
4. Nir Kshetri, *Cybersecurity and Privacy in the Cloud* (MIT Press, 2020).
5. "Zero Trust: Essential for Cloud Security," *Forrester Research Report*, 2021.
6. "Google BeyondCorp: A New Model for Cloud Security," *Google Cloud Blog*, 2020.
7. Alenezi, Mohammad et al., "IAM in Multi-Cloud Security Environments," *Journal of Cloud Security*, 2022.
8. "Amazon GuardDuty: Intelligent Threat Detection in the Cloud," *AWS Blog*, accessed October 9, 2024.
9. "Microsoft Azure Security Center: Advanced Threat Protection," *Microsoft Whitepaper*, 2022.
10. "AWS Compliance Framework: Ensuring Cloud Security Compliance," *AWS Compliance Whitepaper*, 2023.
11. Bisong, Adeshina, and Syed Rahman, *Securing the Cloud: Cloud Computing Security Techniques and Tactics* (Springer, 2021).

2.4 End-to-End Encryption and Secure Communication
References Footnotes

1. Bruce Schneier, *Applied Cryptography: Protocols, Algorithms, and Source Code in C* (Wiley, 2015).
2. "WhatsApp Introduces End-to-End Encryption for Messages and Calls," *WhatsApp Blog*, April 5, 2016.
3. Riana Pfefferkorn, "The Case for End-to-End Encryption: Privacy and Security in the Digital Age," *Stanford Law Review*, 2018.
4. "TLS and Secure Online Banking," *Journal of Financial Security*, 2019.
5. Nir Kshetri, *Cybersecurity and Privacy in the Cloud* (MIT Press, 2020).
6. "Zoom Implements End-to-End Encryption for All Users," *Zoom Blog*, October 27, 2020.

7.   Liu, Jian et al., "End-to-End Encryption in Corporate Communication: A Security Perspective," *IEEE Access*, 2021.
8.   Steven Levy and Bruce Schneier, "The Debate over Encryption Backdoors: Balancing Privacy and Security," *Wired*, September 2020.
9.   Gorjan Alagic and Serge Fehr, "Post-Quantum Cryptography: Ensuring Future-Proof Encryption," *Journal of Cryptographic Research*, 2022.

2.5 Biometrics and Multi-Factor Authentication
References Footnotes

1.   Anil K. Jain, Arun Ross, and Salil Prabhakar, "An Introduction to Biometric Recognition," *IEEE Transactions on Circuits and Systems for Video Technology*, 2004.
2.   "Apple Face ID and Touch ID: Biometric Authentication for the Future," *Apple Developer Documentation*, accessed October 10, 2024.
3.   "HSBC Voice Recognition: Securing Financial Transactions with Biometrics," *HSBC Security Blog*, 2022.
4.   Wouter Gevers and Bart Jacobs, "Multi-Factor Authentication: Enhancing Security Through Layered Defense," *Journal of Network Security*, 2020.
5.   "Time-Based One-Time Passwords: A Secure Alternative to SMS Authentication," *Google Security Blog*, 2019.
6.   "Mastercard Introduces Identity Check for Online Payments," *Mastercard Newsroom*, October 2018.
7.   "Microsoft Azure Active Directory: MFA and Biometric Integration," *Microsoft Security Whitepaper*, 2021.
8.   "2021 Data Breach Investigations Report," Verizon, accessed October 10, 2024.
9.   "HIPAA Compliance and Authentication Security," *Healthcare IT Security Journal*, 2020.
10.   "Researchers Bypass Facial Recognition with 3D-Printed Masks," *IEEE Spectrum*, December 2019.
11.   "Behavioral Biometrics: The Future of Continuous Authentication," *Journal of Cybersecurity Innovation*, 2022.
12.   Alzubaidi, Abdulrahman, and Jugal Kalita, "Biometric Cryptography: Protecting Biometric Data with Encryption," *Journal of Information Security Research*, 2021.

2.6 Quantum Cryptography and Its Potential
References Footnotes

1.   Valerio Scarani et al., "The Security of Practical Quantum Key Distribution," *Reviews of Modern Physics*, 2009.
2.   Michele Mosca, "Cybersecurity in a Quantum World: Will We Be Ready?," *MIT Technology Review*, 2018.
3.   "Swiss Government Deploys Quantum Key Distribution to Secure Communications," *Swiss Cybersecurity Journal*, 2021.
4.   "China Builds World's Longest Quantum Communication Network," *Financial Times*, October 2017.
5.   "Micius: The World's First Quantum Communications Satellite," *Nature*, August 2017.
6.   Charles Bennett and Gilles Brassard, "Quantum Cryptography: Public Key Distribution and Coin Tossing," *Proceedings of IEEE International Conference on Computers, Systems, and Signal Processing*, 1984.
7.   "Quantum-Secure Blockchains: Integrating QKD with Distributed Ledger Technology," *Journal of Cryptographic Research*, 2020.

CHAPTER 3
THE CYBER THREAT LANDSCAPE

3.1 Evolution of Malware: Viruses, Worms, and Trojans
References Footnotes

1.   "Creeper: The First Computer Virus," *Journal of Cybersecurity History*, 2018.
2.   "The Melissa Virus: A Case Study in Macro Viruses," *Journal of Information Security*, 2001.
3.   Gordon, Lawrence A., and Martin P. Loeb, *Managing Cybersecurity Resources: A Cost-Benefit Analysis* (McGraw-Hill, 2006).
4.   "The Morris Worm: Lessons from the First Major Internet Attack," *IEEE Security and Privacy*, 2013.
5.   "ILOVEYOU: The Virus That Crippled the Internet," *BBC News*, May 2000.
6.   "Stuxnet: The First Weaponized Malware," *Wired*, 2011.
7.   Thomas Rid and Ben Buchanan, "Attributing Cyber Attacks," *Journal of Strategic Studies*, 2015.
8.   "The AIDS Trojan: The First Ransomware," *Journal of Computer Security*, 1991.
9.   "Zeus Trojan: A Case Study in Cybercrime," *Forbes*, 2010.
10.   Kim-Kwang Raymond Choo, "The Cyber Threat Landscape: Challenges and Future Research Directions," *Computers & Security*, 2011.
11.   Bruce Schneier, *Secrets and Lies: Digital Security in a Networked World* (Wiley, 2015).

3.2 Ransomware Attacks and Their Economic Impact
References Footnotes

1. "WannaCry: The Ransomware Attack That Shook the World," *BBC News*, May 2017.
2. Andy Greenberg, *Sandworm: A New Era of Cyberwar and the Hunt for the Kremlin's Most Dangerous Hackers* (Doubleday, 2019).
3. "Colonial Pipeline Ransomware Attack: Lessons Learned," *The Washington Post*, May 2021.
4. "The State of Ransomware 2021: Sophos Report," *Sophos*, accessed October 10, 2024.
5. "WannaCry's Impact on the UK's NHS: Economic and Operational Losses," *Journal of Healthcare IT*, 2018.
6. M. Van Eeten, "Ransomware and the Economics of Downtime," *Journal of Cybersecurity*, 2020.
7. "Norsk Hydro Ransomware Attack: Financial Impact and Recovery," *Forbes*, April 2019.
8. "CWT Global Ransomware Attack Compromises Employee Data," *Reuters*, July 2020.
9. Mansoor Kharraz and William Robertson, "The Rise of Ransomware-as-a-Service: Implications for Cybersecurity," *Journal of Computer Security*, 2021.
10. "Global Ransomware Damage Costs Expected to Reach $20 Billion in 2021," *Cybersecurity Ventures*, 2021.
11. "Ransomware Attacks Disrupt Education Sector," *The New York Times*, September 2021.
12. "Ransomware Threats to U.S. Critical Infrastructure," *Department of Homeland Security Report*, 2021.
13. "Cyber Insurance and Ransomware: Navigating the Risks," *Journal of Insurance and Risk Management*, 2022.

3.3 Social Engineering and Phishing Techniques
References Footnotes

1. Kevin Mitnick and William Simon, *The Art of Deception: Controlling the Human Element of Security* (Wiley, 2002).
2. "The RSA Security Breach: Lessons in Social Engineering," *Journal of Cybersecurity*, 2012.
3. "The Rise of Phishing: How PayPal Became a Target," *Computer Fraud & Security*, 2004.
4. "The John Podesta Email Hack: Spear Phishing in U.S. Politics," *The New York Times*, October 2016.
5. Thomas Jagatic et al., "Social Phishing: Leveraging Social Networks in Phishing Attacks," *Communications of the ACM*, 2007.
6. "Whaling Attack Costs Ubiquiti Networks $46 Million," *Forbes*, March 2016.
7. Katherine Parsons et al., "The Psychology of Spear Phishing," *Journal of Cybersecurity*, 2020.
8. "Twitter Vishing Attack: How Hackers Took Over Verified Accounts," *The Verge*, July 2020.
9. Jason Hong, "The State of Phishing: Emerging Threats in Vishing and Smishing," *IEEE Security & Privacy*, 2021.
10. "FBI Reports Record Losses from Business Email Compromise in 2020," *FBI Internet Crime Report*, 2021.
11. "Toyota Supplier Hit by $37 Million Business Email Compromise Scam," *Bloomberg*, 2019.
12. Alice Hutchings and Richard Clayton, "Exploring the Financial Impact of Business Email Compromise," *Journal of Financial Crime*, 2021.
13. "2021 Data Breach Investigations Report," Verizon, accessed October 10, 2024.

3.4 Advanced Persistent Threats (APTs) and Nation-State Actors
References Footnotes

1. Thomas Rid and Ben Buchanan, "Attributing Cyber Attacks," *Journal of Strategic Studies*, 2015.
2. "APT28: A Detailed Look at the Russian Cyber Espionage Group," *FireEye Threat Intelligence Report*, 2018.
3. "APT29 Targeting COVID-19 Vaccine Research," *U.S. National Cyber Security Centre (NCSC)*, July 2020.
4. "APT10: Chinese Cyber Espionage Campaign Targeting Global Trade," *Journal of International Security*, 2018.
5. Alissa Zander et al., "The Tactics, Techniques, and Procedures of Advanced Persistent Threats," *IEEE Security & Privacy*, 2019.
6. "Operation Aurora: A Sophisticated Cyber Espionage Attack," *Google Security Blog*, 2010.
7. "Chinese APT Groups Targeting Intellectual Property," *Journal of Cybersecurity Research*, 2021.
8. Kim Zetter, *Countdown to Zero Day: Stuxnet and the Launch of the World's First Digital Weapon* (Crown, 2014).
9. Richard A. Clarke and Robert Knake, *Cyber War: The Next Threat to National Security and What to Do About It* (HarperCollins, 2010).
10. "Defending Against APTs: Lessons from the Frontlines," *FireEye Mandiant Report*, 2022.

3.5 Insider Threats and Privileged Access Management
References Footnotes

1. Linda Greitzer et al., "Combating the Insider Cyber Threat," *Journal of Information Warfare*, 2014.
2. Edward Snowden, *Permanent Record* (Metropolitan Books, 2019).
3. Dawn Cappelli et al., *The CERT Guide to Insider Threats* (Addison-Wesley, 2012).
4. "2020 Cost of Insider Threats Global Report," *Ponemon Institute*, 2020.

5. "Privileged Access Management: Market Guide," *Gartner Research*, 2021.
6. "The Target Breach: Lessons in Privileged Access Management," *CSO Magazine*, March 2014.
7. Elisa Bertino and Ravi Sandhu, "Privileged Access Management: Concepts and Best Practices," *IEEE Security and Privacy*, 2019.
8. "Centrify: Just-in-Time Privileged Access Management," *Centrify Whitepaper*, 2021.
9. Iman Gheyas and Samer Abdallah, "AI and Machine Learning in Insider Threat Detection," *Journal of Cybersecurity Research*, 2020

3.6 Emerging Threats in the IoT and Mobile Landscapes
References Footnotes

1. "The Growth of IoT: Trends and Predictions," *Forbes*, 2022.
2. Yang, Haibo et al., "Security in the Internet of Things: Vulnerabilities and Solutions," *IEEE Internet of Things Journal*, 2019.
3. "The Mirai Botnet Attack: A Wake-Up Call for IoT Security," *Wired*, November 2016.
4. "Smart Home Systems Vulnerable to Hijacking," *Journal of Cybersecurity Research*, 2020.
5. Lee, Kwang et al., "Securing Critical Infrastructure in the IoT Era," *Journal of Industrial Security*, 2021.
6. "Ponemon Institute IoT Security Study," *Ponemon Institute Report*, 2020.
7. "Mobile Security Threats in 2021: Check Point Research Report," *Check Point Research*, accessed October 11, 2024.
8. "Mobile Phishing Attacks on the Rise: Lookout Report," *Lookout Security Blog*, 2021.
9. Sinha, Anurag, "The Role of Mobile Device Management in Enterprise Security," *Journal of Enterprise Security*, 2019.
10. Meidan, Yair et al., "The Security Challenges of 5G-Enabled Smart Cities," *IEEE Communications Magazine*, 2021.
11. Jain, Shikhar et al., "Best Practices for Mobile Security in the Enterprise," *Journal of Mobile Security*, 2020.

CHAPTER 4
CYBERSECURITY FOR INDIVIDUALS

4.1 Personal Data Protection Strategies
References Footnotes

1. "2020 Identity Theft and Fraud Report," *Federal Trade Commission (FTC)*, accessed October 11, 2024.
2. Barton Gellman, *Dark Mirror: Edward Snowden and the American Surveillance State* (Penguin Press, 2020).
3. "2021 Data Breach Investigations Report," Verizon, accessed October 11, 2024.
4. "Twitter Breach: A Lesson in Multi-Factor Authentication," *The New York Times*, July 2020.
5. "End-to-End Encryption: Securing Messages with WhatsApp and iMessage," *Journal of Communication Security*, 2020.
6. "The Cost of Unpatched Vulnerabilities," *Ponemon Institute Report*, 2021.
7. "Ransomware: How Backups Can Save Your Data," *Kaspersky Lab Whitepaper*, 2022.
8. Bruce Schneier, *Data and Goliath: The Hidden Battles to Collect Your Data and Control Your World* (W.W. Norton, 2015).
9. "Google Drive Phishing Attack Targets Millions," *Wired*, May 2016.

4.2 Secure Online Behavior and Digital Hygiene
References Footnotes

1. Lawrence A. Gordon and Martin P. Loeb, *Managing Cybersecurity Resources: A Cost-Benefit Analysis* (McGraw-Hill, 2006).
2. Bruce Schneier, *Data and Goliath: The Hidden Battles to Collect Your Data and Control Your World* (W.W. Norton, 2015).
3. "Public Wi-Fi Security Risks and Best Practices," *Norton Security Blog*, 2020.
4. Ross Anderson and Tyler Moore, "Cybersecurity Economics: Lessons Learned and Future Challenges," *IEEE Security & Privacy*, 2018.

4.3 Password Management and Authentication Best Practices
References Footnotes

1. Dinei Florencio and Cormac Herley, "A Large-Scale Study of Web Password Habits," *Proceedings of the International World Wide Web Conference*, 2007.
2. "The 2020 Worst Passwords List," *SplashData*, 2020.
3. "2021 Data Breach Investigations Report," Verizon, accessed October 12, 2024.
4. Ross Anderson and Tyler Moore, "The Economics of Information Security," *Science*, 2006.
5. "Google Cloud Platform Security Breach: The Role of MFA," *Google Security Blog*, 2019.

6. Florencio and Herley, "A Study of Web Password Habits," 2007.
7. "Apple Face ID and Touch ID: Biometric Authentication for Secure Devices," *Apple Developer Documentation*, accessed October 12, 2024.
8. Anil Jain et al., "Biometric Authentication: Security and Privacy Concerns," *IEEE Security & Privacy*, 2016.
9. Frank Stajano and Ross Anderson, "The Resurgence of Passwordless Authentication," *IEEE Symposium on Security and Privacy*, 2021.

4.4 Protecting Personal Devices and Home Networks
References Footnotes

1. "Internet Security Threat Report," *Symantec*, 2020.
2. "The 2020 Ponemon Institute Report on Cybersecurity Risks," *Ponemon Institute*, 2020.
3. "Encryption in Apple and Android Devices: A Comparison," *Journal of Mobile Security*, 2019.
4. "Router Security Best Practices," *Kaspersky Lab*, 2021. "IoT Device Growth and Security Challenges," *Gartner*, 2022.
5. "The Role of VPNs in Personal Network Security," *NordVPN Whitepaper*, 2021.
6. "Parental Controls and Internet Safety for Families," *Common Sense Media*, 2020.
7. Anurag Sinha and Mohan Rao, "The Future of AI in Home Network Security," *IEEE Security and Privacy*, 2022.

4.6 Recognizing and Avoiding Common Cyber Scams
References Footnotes

1. "Phishing Attack Trends: Annual Report," *Anti-Phishing Working Group (APWG)*, 2022.
2. "Google Drive Phishing Scam: 2016 Incident," *The New York Times*, May 2016.
3. "2020 Consumer Sentinel Network Data Book," *Federal Trade Commission (FTC)*, accessed October 12, 2024.
4. "Tech Support Scams: A Growing Threat," *Microsoft Security Blog*, 2018.
5. "Avoiding Online Shopping Scams," *Norton Security Insights*, 2021.
6. "Surge in Fake E-commerce Sites During Pandemic," *Journal of Consumer Protection*, 2020.
7. "Romance Scams and Financial Exploitation," *FBI Internet Crime Complaint Center (IC3) Report*, 2021.
8. "Cryptocurrency Scams: 2021 Insights," *Chainalysis*, 2022.
9. Kevin Mitnick and William Simon, *The Art of Deception: Controlling the Human Element of Security* (Wiley, 2002).

CHAPTER 5
ORGANIZATIONAL CYBERSECURITY

5.1 Develop a Comprehensive Cybersecurity Strategy
References Footnotes

1. "The Cost of Data Breaches: 2022 Insights," *Gartner Research*, accessed October 12, 2024.
2. "Cybersecurity Framework," *National Institute of Standards and Technology (NIST)*, 2021.
3. "Information Security Management Systems: ISO/IEC 27001 Standard," *ISO/IEC*, 2020.
4. Alok Saxena et al., "AI and Machine Learning in Cybersecurity: Challenges and Opportunities," *Journal of Cybersecurity Research*, 2021.
5. "Incident Response: Best Practices," *SANS Institute*, 2020.
6. "2021 Data Breach Investigations Report," *Verizon*, accessed October 12, 2024.
7. Jay P. Kesan and Carol M. Hayes, "Cybersecurity and Privacy Regulations: Ensuring Compliance in a Changing Landscape," *Journal of Law and Technology*, 2019.

5.2 Risk Assessment and Management Frameworks
References Footnotes

1. "Risk Management Framework," *National Institute of Standards and Technology (NIST)*, 2020.
2. "NIST Cybersecurity Framework," *NIST*, accessed October 12, 2024.
3. "ISO 31000: Risk Management Guidelines," *International Organization for Standardization (ISO)*, 2018.
4. "The Role of Cyber Insurance in Risk Management," *Marsh & McLennan*, 2021.
5. Thomas R. Peltier, *Information Security Risk Analysis* (CRC Press, 2013).
6. Jay P. Kesan and Carol M. Hayes, "Compliance and Cybersecurity Regulations: Challenges and Opportunities," *Journal of Law and Technology*, 2019.
7. "Artificial Intelligence in Risk Management," *PwC*, 2022.

5.3 Implement Security
References Footnotes

1. "Information Security Guidelines: ISO/IEC 27002," *International Organization for Standardization (ISO)*, 2021.
2. "Risk-Based Approach to Cybersecurity Policy Development," *National Institute of Standards and Technology (NIST)*, 2020.
3. Jay P. Kesan and Carol M. Hayes, "Cybersecurity Compliance: Legal and Regulatory Challenges," *Journal of Cybersecurity Law*, 2019.
4. "Building a Culture of Security Awareness," *SANS Institute Report*, 2020.
5. "Zero Trust Security: Adapting to Remote Work Challenges," *Forrester Research*, 2021.
6. Elisa Bertino and Ravi Sandhu, "Information Security Policies: Best Practices and Guidelines," *IEEE Security and Privacy*, 2019.

5.4 Incident Response Planning and Execution
References Footnotes

1. "Incident Response Planning: A Comprehensive Approach," *ENISA*, 2021.
2. "Incident Response Best Practices," *SANS Institute*, 2020.
3. "NIST Cybersecurity Framework: Incident Response Guidance," *National Institute of Standards and Technology (NIST)*, 2020.
4. "The SolarWinds Hack: A Case Study in Incident Response," *Cybersecurity Journal*, 2021.
5. "2021 Data Breach Investigations Report," Verizon, accessed October 12, 2024.
6. "Cost of a Data Breach Report," *IBM Security*, 2022.
7. Deborah Bodeau and Robert Graubart, "Improving Incident Response Through Post-Incident Reviews," *Journal of Information Security*, 2019.
8. John Crowe and Susan Riley, *Incident Response Strategies: Coordinating IT, Legal, and PR*, 2020.
9. "Cyber Insurance and Incident Response," *Marsh McLennan*, 2022.

5.5 Business Continuity and Disaster Recovery
References Footnotes

1. "Cost of a Data Breach Report," *IBM Security*, 2022.
2. "Developing Disaster Recovery Strategies for Critical Systems," *Gartner Research*, 2020.
3. "The Role of Business Impact Analysis in BC/DR Planning," *The Business Continuity Institute*, 2021.
4. Lisa Paul and Michael Murry, *Crisis Communications: Managing Public Relations in Times of Crisis*, 2020.
5. "Best Practices for Testing BC/DR Plans," *SANS Institute*, 2021.
6. "Cloud-Based Disaster Recovery Solutions: Benefits and Best Practices," *IDC Research*, 2022.
7. Jay P. Kesan and Carol M. Hayes, "Ensuring Compliance with Regulatory Requirements in BC/DR Plans," *Journal of Cybersecurity Law*, 2019.

5.6 Third-Party Risk Management and Supply Chain Security
References Footnotes

1. "The Rise of Supply Chain Attacks," *KPMG Cybersecurity Report*, 2021.
2. "The SolarWinds Attack: A Case Study in Supply Chain Security," *Journal of Cybersecurity Research*, 2020.
3. "Third-Party Risk Management: Best Practices," *Forrester Research*, 2022.
4. "Lessons from the Target Breach: Managing Third-Party Risks," *CSO Magazine*, 2014.
5. "Supply Chain Security in a Globalized World," *PwC Whitepaper*, 2021.
6. Anurag Sinha et al., "Coordinating Incident Response Across Supply Chains," *Journal of Information Security*, 2021.
7. "Ensuring Compliance in Third-Party Risk Management," *Gartner Research*, 2022.
8. Yair Meidan et al., "AI and Blockchain in Supply Chain Security," *IEEE Security & Privacy*, 2020.

CHAPTER 6
CYBERSECURITY IN KEY INDUSTRIES

6.1 Financial Services: Protect Sensitive Transactions and Data
References Footnotes

1. "2021 Cybercrime Report," *Accenture*, accessed October 12, 2024.
2. "CNA Financial Ransomware Attack," *Cybersecurity Journal*, 2020.
3. "Tokenization in Payment Security," *PCI Security Standards Council*, 2021.
4. "2021 Data Breach Investigations Report," *Verizon*, accessed October 12, 2024.
5. "PCI DSS Compliance Guidelines," *PCI Security Standards Council*, 2020.
6. Jay P. Kesan and Carol M. Hayes, "Data Protection and Privacy: The Role of GDPR in Financial Services," *Journal of Law and Technology*, 2019.
7. "AI in Fraud Detection: Trends and Insights," *PwC Report*, 2022.
8. "Incident Response Best Practices," *SANS Institute*, 2021.
9. "Cyber Resilience in Financial Services," *Deloitte Insights*, 2022.

6.2 Healthcare: Secure Patient Information and Medical Devices
References Footnotes

1. "Healthcare Data Breach Report," *Protenus*, 2022.
2. "UHS Hospitals Ransomware Attack: A Case Study," *Healthcare Cybersecurity Journal*, 2020.
3. "2021 Data Breach Investigations Report," *Verizon*, accessed October 12, 2024.
4. "FDA Report on Pacemaker Vulnerabilities," *U.S. Food and Drug Administration (FDA)*, 2017.
5. Jay P. Kesan and Carol M. Hayes, "Data Protection in Healthcare: Navigating HIPAA and GDPR Compliance," *Journal of Cybersecurity Law*, 2019.
6. "Multi-Factor Authentication in Healthcare: Trends and Insights," *Deloitte Cybersecurity Report*, 2021.
7. "Incident Response in Healthcare: Best Practices," *SANS Institute*, 2020.
8. "Blockchain in Healthcare: Enhancing Cybersecurity and Privacy," *PwC Whitepaper*, 2022.

6.3 Energy and Utilities: Safeguarding Critical Infrastructure
References Footnotes

1. "The State of Industrial Control Systems Security," *Dragos Report*, 2022.
2. "The Ukraine Power Grid Hack: A Case Study in Cybersecurity," *Cybersecurity Journal*, 2015.
3. "NIST Cybersecurity Framework for Industrial Control Systems," *National Institute of Standards and Technology (NIST)*, 2021.
4. "Colonial Pipeline Ransomware Attack: Lessons Learned," *Journal of Energy Security*, 2021.
5. "Supply Chain Security in the Energy Sector," *PwC Whitepaper*, 2021.
6. "The Impact of the NIS Directive on Critical Infrastructure Security," *ENISA*, 2020.
7. "Real-Time Threat Detection and Response in Critical Infrastructure," *SANS Institute*, 2021.
8. "Cybersecurity for Renewable Energy Systems: Trends and Best Practices," *Deloitte Insights*, 2021.

6.4 Retail and E-Commerce: Ensuring Secure Transactions and Customer Trust
References Footnotes

1. "Retail Threat Landscape Report," *Fortinet*, 2021.
2. "Target Data Breach: Lessons Learned," *Journal of Cybersecurity and Privacy*, 2014.
3. "Tokenization and Payment Security in Retail," *PCI Security Standards Council*, 2021.
4. "2021 Data Breach Investigations Report," *Verizon*, accessed October 12, 2024.
5. "Fraud Detection in Retail: Best Practices and Trends," *Deloitte*, 2022.
6. "Vulnerability Scanning and Patch Management Guidelines," *National Institute of Standards and Technology (NIST)*, 2021.
7. "Building Customer Trust in Retail: The Role of Privacy," *PwC Retail Insights*, 2021.
8. "Cyber Resilience in Retail: Best Practices," *SANS Institute*, 2020.

6.5 Educational Institutions: Balancing Openness with Data Protection
References Footnotes

1. "2022 Data Breach Investigations Report," *Verizon*, accessed October 12, 2024.
2. "Ransomware Attack on Newhall School District," *Education Cybersecurity Journal*, 2020.
3. "Student Data Privacy and Cybersecurity Best Practices," *Educause*, 2021.
4. Rick Schatz et al., "Balancing Openness and Security in Higher Education," *Journal of Information Security in Academia*, 2020.
5. "Michigan State University Ransomware Attack: A Case Study," *Journal of Higher Education Cybersecurity*, 2020.
6. "Multi-Factor Authentication in Education: Reducing the Risk of Cyberattacks," *Deloitte Cybersecurity Report*, 2021.
7. "Securing Online Learning Platforms: Best Practices for Schools," *National Cyber Security Centre (NCSC)*, 2020.
8. "Cyber Resilience in Education: Strategies for Recovery," *SANS Institute*, 2021.

6.6 Manufacturing: Securing Smart Factories and Industrial Control Systems
References Footnotes

1. "Securing Smart Factories in the Digital Age," *McKinsey Insights*, 2022.
2. "State of Industrial Control System Security," *Dragos*, 2022.
3. "Industrial IoT: Cybersecurity Risks and Mitigation Strategies," *Gartner Research*, 2021.
4. "IEC 62443 Standard for Industrial Control Systems Security," *International Electrotechnical Commission (IEC)*, 2021.
5. "Ransomware Attack on Norsk Hydro: Impact and Lessons Learned," *Journal of Industrial Security*, 2019.
6. "Incident Response in Manufacturing: Best Practices," *SANS Institute*, 2021.
7. "Supply Chain Attack on SolarWinds: A Case Study," *Journal of Cybersecurity Research*, 2020.

CHAPTER 7
THE ECONOMICS OF CYBERSECURITY

7.1 The Cost of Cyber Attacks on Businesses and Economies
References Footnotes

1. "2022 Cost of Cybercrime Study," *Accenture*, accessed October 12, 2024.
2. "Cost of a Data Breach Report," *IBM Security*, 2021.
3. "The 2021 Ransomware Report," *Sophos*, accessed October 12, 2024.
4. "Colonial Pipeline Ransomware Attack: Financial and Economic Impact," *Journal of Cybersecurity and Critical Infrastructure*, 2021.
5. "Economic Impact of Cyberattacks on Critical Infrastructure," *ENISA*, 2020.
6. "Economic Impact of Cybercrime Report," *McAfee*, 2020.
7. "British Airways Fined £20 Million for Data Breach," *BBC News*, 2021.
8. "Equifax Fined $700 Million for 2017 Data Breach," *Reuters*, 2021.
9. "The Rising Cost of Regulatory Compliance," *Deloitte Insights*, 2022.
10. "Cyber Insurance Premiums Surge as Attacks Increase," *Marsh McLennan Cybersecurity Report*, 2021.
11. "Cybersecurity Risks and Economic Stability," *PwC Global Insights*, 2022.

7.2 Cybersecurity Investment Strategies and ROI
References Footnotes

1. "Global Cybersecurity Spending Projections," *Gartner Insights*, 2022.
2. "Cost of a Data Breach Report," *IBM Security*, 2021.
3. "Consumer Trust and Cybersecurity: Insights from the Marketplace," *PwC Survey*, 2021.
4. "The Rise in Cyber Insurance Premiums," *Marsh McLennan Cybersecurity Report*, 2021.
5. "Lowering Cyber Insurance Premiums with Strong Security Measures," *Forrester Research*, 2022.
6. "Aligning Cybersecurity Investment with Business Strategy," *McKinsey Insights*, 2021.

7.3 The Cyber Insurance Market and Risk Transfer
References Footnotes

1. "The Global Cyber Insurance Market: Trends and Projections," *Allied Market Research*, 2022.
2. "The Challenge of Underwriting Cyber Risk," *PwC Insights*, 2021.
3. "The NotPetya Attack: A Case Study in Cyber Insurance Exclusions," *Journal of Information Security*, 2017.
4. "The Future of Cyber Insurance: Regulation and Standardization," *Deloitte Insights*, 2022.

7.4 Economic Incentives for Improving Cybersecurity
References Footnotes

1. "Cybersecurity Tax Credits and Incentives," *PwC Insights*, 2021.
2. "Horizon 2020: Cybersecurity Research and Funding," *European Union Cybersecurity Initiative*, 2022.
3. "Cyber Insurance Premium Discounts for Strong Security Measures," *Marsh McLennan*, 2021.
4. "Public-Private Partnerships in Cybersecurity: Industry 100 Initiative," *National Cyber Security Centre (NCSC)*, 2020.
5. "GDPR Compliance and Economic Incentives," *Journal of Data Protection and Privacy*, 2021.
6. "DARPA's Role in Cybersecurity Research and Innovation," *DARPA Report*, 2021.

7.5 The Cybersecurity Job Market and Skills Gap
References Footnotes

1. "Global Cybersecurity Workforce Shortage," *Cybersecurity Ventures*, 2022.
2. "(ISC)² Cybersecurity Workforce Study," *(ISC)² Report*, 2021.
3. "Closing the Cybersecurity Skills Gap," *Fortinet Cybersecurity Report*, 2022.
4. "The Impact of the Cybersecurity Skills Shortage," *Accenture Insights*, 2021.
5. "Cybersecurity Internships and Apprenticeships: A Path to Closing the Skills Gap," *Deloitte Insights*, 2021.
6. "The Role of Certifications in Cybersecurity," *Gartner Research*, 2021.
7. "Cybersecurity Scholarships and Grants," *U.S. Department of Homeland Security (DHS)*, 2021.

7.6 Cybercrime Economics and the Dark Web Marketplace
References Footnotes

1. "The Evolution of the Dark Web Marketplace," *The Rand Corporation*, 2021.
2. "The Economic Impact of Cybercrime," *McAfee*, 2020.
3. "Ransomware Report 2021," *Sophos*, accessed October 12, 2024.
4. "Crypto Crime Report," *Chainalysis*, 2021.
5. "The Rise of Ransomware as a Service," *Check Point Research*, 2021.
6. "International Efforts to Combat Cybercrime on the Dark Web," *Interpol*, 2021.

CHAPTER 8
ETHICAL AND LEGAL ASPECTS OF CYBERSECURITY

8.1 Privacy vs. Security: Striking the Right Balance
References Footnotes

1. "2022 Cybercrime Report," *Accenture*, accessed October 12, 2024.
2. Daniel J. Solove and Paul M. Schwartz, "Privacy, Security, and the Balance: Finding the Ethical Middle Ground," *Journal of Law and Technology*, 2019.
3. "The San Bernardino Case: Encryption and Privacy in the Spotlight," *New York Times*, 2016.
4. "Surveillance and Privacy Concerns," *Electronic Frontier Foundation (EFF)*, 2021.
5. "EU Court Strikes Down Privacy Shield: Implications for Global Data Transfers," *European Court of Justice*, 2020.
6. Ann Cavoukian and Jeff Jonas, "Privacy by Design: Achieving Both Privacy and Security," *Journal of Information Security*, 2020.

8.2 Ethical Hacking and Responsible Disclosure
References Footnotes

1. "Global Demand for Ethical Hackers and the Growing Skills Gap," *Cybersecurity Ventures*, 2022.
2. "The Role of Ethical Hacking in Cybersecurity," *McKinsey Insights*, 2021.
3. Bruce Schneier, "Responsible Disclosure: The Ethics of Reporting Vulnerabilities," *Journal of Information Security*, 2020.
4. "The Importance of CERT/CC in Coordinating Responsible Disclosure," *CERT/CC Annual Report*, 2021.
5. Kevin Mitnick, "Legal Challenges in Ethical Hacking," *Journal of Cyber Law*, 2020.
6. "Bug Bounty Programs and Their Impact on Cybersecurity," *HackerOne*, 2022.
7. "The Impact of Ethical Hacking on Public Trust," *Forrester Research*, 2021.
8. "Ethical Hacking and the Protection of Critical Infrastructure," *Journal of Critical Infrastructure Protection*, 2017.

8.3 Government Surveillance and Civil Liberties
References Footnotes

1.   "The Role of Government Surveillance in National Security," *Journal of Homeland Security*, 2021.
2.   Daniel J. Solove, "The Ethics of Privacy in the Age of Mass Surveillance," *Journal of Law and Technology*, 2019.
3.   James Bamford, "Mass Surveillance and the Erosion of Privacy Rights," *Journal of Civil Liberties*, 2020.
4.   Jonathan Penney, "The Chilling Effect of Government Surveillance on Free Speech," *Journal of Human Rights and Technology*, 2021.
5.   Glenn Greenwald, "Oversight and Accountability in Government Surveillance," *Journal of National Security Law*, 2020.

8.4 Corporate Responsibility in Data Breaches
References Footnotes

1.   "Equifax Data Breach Settlement," *U.S. Federal Trade Commission (FTC)*, 2019.
2.   Daniel J. Solove, "Ethical Obligations in Data Protection," *Journal of Law and Technology*, 2020.
3.   "GDPR and Data Breach Reporting," *European Commission*, 2021.
4.   "British Airways Fined £20 Million for Data Breach," *UK Information Commissioner's Office (ICO)*, 2021.
5.   "Post-Breach Response: Transparency and Accountability," *McKinsey Insights*, 2021.
6.   "The Role of Corporate Culture in Data Protection," *Deloitte Cybersecurity Report*, 2022.
7.   "Yahoo Data Breach Settlement," *Reuters*, 2020.

8.5 International Laws and Regulations on Cybersecurity
References Footnotes

1.   "The Role of ENISA in International Cybersecurity," *European Union Agency for Cybersecurity (ENISA)*, 2021.
2.   "General Data Protection Regulation (GDPR): A Global Standard," *European Commission*, 2020.
3.   "Budapest Convention on Cybercrime," *Council of Europe*, accessed October 12, 2024.
4.   "China's Cybersecurity Law: National Security and Data Control," *Journal of Asian Law and Policy*, 2019.
5.   Michael Schmitt, "Challenges in International Cybersecurity Law," *Journal of International Law and Technology*, 2020.
6.   "The Global Cybersecurity Agenda (GCA): ITU's Framework for International Cooperation," *International Telecommunication Union (ITU)*, 2021.
7.   "The African Union's Malabo Convention on Cybersecurity," *African Union*, 2020.
8.   Melissa Hathaway, "The Role of Cyber Diplomacy in International Relations," *Journal of Cybersecurity Studies*, 2021.
9.   "The EU's Cyber Diplomacy Toolbox," *European Union Cybersecurity Strategy*, 2021.

8.6 Ethical Considerations in AI-Driven Security Systems
References Footnotes

1.   "AI and the Future of Cybersecurity," *Gartner Research*, 2021.
2.   Reuben Binns, "Algorithmic Bias in AI Systems: Ethical Considerations," *Journal of Information Security Ethics*, 2020.
3.   Shoshana Zuboff, "Surveillance Capitalism and the Threat to Privacy in the Age of AI," *Journal of Human Rights and Technology*, 2019.
4.   Frank Pasquale, "The Black Box Society: The Secret Algorithms That Control Decisions," *Journal of Law and Technology*, 2018.
5.   Iyad Rahwan, "Autonomy and Over-Reliance on AI: Implications for Cybersecurity," *Journal of AI and Society*, 2021.
6.   "Ethics Guidelines for Trustworthy AI," *European Commission*, 2020.

CHAPTER 9
GLOBAL CYBERSECURITY LANDSCAPE

9.1 Cyber Warfare and Nation-State Activities
References Footnotes

1.   Richard A. Clarke and Robert Knake, "Cyber War: The Next Threat to National Security," *Journal of Strategic Studies*, 2019.
2.   Kim Zetter, "Countdown to Zero Day: Stuxnet and the Launch of the World's First Digital Weapon," *Journal of Cybersecurity Studies*, 2018.
3.   "The NotPetya Cyberattack and Its Global Impact," *Journal of Information Security*, 2018.
4.   Thomas Rid, "Cyber Espionage and the Role of Nation-States," *Journal of Intelligence and National Security*, 2020.
5.   "Russia's Role in the 2016 U.S. Election: A Cyber Perspective," *Journal of International Security Studies*, 2017.

6. Thomas Rid and Ben Buchanan, "Attribution in Cyber Warfare: Challenges and Strategies," *Journal of Strategic Studies*, 2021.
7. "NATO's Cyber Defense Strategy: Securing the Digital Frontier," *NATO Review*, 2020.
8. Michael N. Schmitt, "The Future of Cyber Warfare: Law and Strategy," *Journal of National Security Law*, 2020.

9.2 International Cooperation in Fighting Cybercrime
References Footnotes

1. "Interpol's Role in Combating Cybercrime," *Interpol Global Cybercrime Programme*, 2021.
2. "Budapest Convention on Cybercrime," *Council of Europe*, accessed October 12, 2024.
3. "ENISA's Report on the Budapest Convention," *European Union Agency for Cybersecurity (ENISA)*, 2020.
4. "The African Union's Malabo Convention on Cybersecurity," *African Union*, 2020.
5. "Europol's European Cybercrime Centre: Achievements and Operations," *Europol Annual Report*, 2021.
6. "Operation GOLDFISH ALPHA: An International Ransomware Takedown," *Europol Cybercrime Operations*, 2021.
7. Michael Schmitt, "Geopolitical Challenges in International Cybersecurity Cooperation," *Journal of Strategic Cybersecurity*, 2020.
8. "Global Programme on Cybercrime," *United Nations Office on Drugs and Crime (UNODC)*, 2021.
9. Thomas Rid, "Norms and State Behavior in Cyberspace," *Journal of International Security Studies*, 2021.

9.3 Global Cybersecurity Treaties and Agreements
References Footnotes

1. "ENISA's Role in Global Cybersecurity Cooperation," *European Union Agency for Cybersecurity (ENISA)*, 2021.
2. Michael Schmitt, "The Tallinn Manual and the Legal Framework for Cyber Warfare," *Journal of International Law and Security*, 2021.
3. "UN Group of Governmental Experts (GGE) Report on Cybersecurity," *United Nations*, 2021.
4. "Public-Private Partnerships in Global Cybersecurity," *Deloitte Cybersecurity Insights*, 2022.

9.4 Comparative Analysis of National Cybersecurity Strategies
References Footnotes

1. "National Cyber Strategy of the United States," *White House Cybersecurity Report*, 2021.
2. "General Data Protection Regulation and the EU Cybersecurity Act," *ENISA Report*, 2021.
3. Ronald Deibert, "China's Cyber Sovereignty and the Role of the State," *Journal of International Affairs*, 2020.
4. Thomas Rid, "Cyber Warfare and Russia's Geopolitical Strategy," *Journal of Strategic Studies*, 2019.
5. "Japan's Cybersecurity Strategy: Securing Critical Infrastructure," *Ministry of Internal Affairs and Communications Report*, 2020.

9.5 Cross-Border Data Flow and Jurisdiction Issues
References Footnotes

1. "Cross-Border Data Flows and Global Economic Growth," *McKinsey Global Institute*, 2020.
2. "GDPR and Cross-Border Data Transfers," *European Commission*, 2021.
3. "The Invalidity of the Privacy Shield: Implications for Data Transfers," *Journal of Data Protection Law*, 2020.
4. Ronald Deibert, "Data Sovereignty and National Security in the Digital Age," *Journal of International Affairs*, 2019.
5. "The CLOUD Act: Cross-Border Data Access and Law Enforcement," *Journal of Cybersecurity Law*, 2018.
6. "Privacy Concerns in the CLOUD Act," *Human Rights Watch*, 2019.
7. "Budapest Convention on Cybercrime: International Cooperation on Data Protection," *Council of Europe*, 2021.

9.6 Building a Global Culture of Cybersecurity
References Footnotes

1. David Schatz and Rashid Bashroush, "A Collective Approach to Building a Global Cybersecurity Culture," *Journal of Cybersecurity Policy*, 2020.
2. Morgan, Steve, and Bada, Maria, "The Role of Education in Combatting Global Cyber Threats," *Journal of Cybersecurity Education*, 2019.
3. "International Standards for Cybersecurity," *ISO/IEC Standards*, 2021.
4. "The Digital Divide in Global Cybersecurity Readiness," *ITU Cybersecurity Report*, 2020.

CHAPTER 10
THE FUTURE OF CYBERSECURITY

10.1 Quantum Computing: Threat and Opportunity
References Footnotes

1. Michele Mosca, "The Quantum Threat to Cryptography: Preparing for the Future," *Journal of Cryptography and Information Security*, 2020.
2. "Quantum Computing and the Future of Encryption," *NIST Report on Post-Quantum Cryptography*, 2022.
3. Lintao Chen and Stephen Jordan, "The Race to Build Quantum Computers: Implications for Cybersecurity," *Journal of Emerging Technologies*, 2021.
4. "Post-Quantum Cryptography Standardization: NIST's Role in Preparing for the Quantum Threat," *NIST Post-Quantum Cryptography Project*, 2022.
5. Charles Bennett and Gilles Brassard, "Quantum Key Distribution: The Future of Secure Communication," *Journal of Quantum Information Science*, 2018.
6. "China's Quantum Leap: The Launch of the Micius Satellite and Its Implications for Global Security," *Journal of Quantum Technologies*, 2017.
7. "Quantum Technologies and Global Competitiveness: An OECD Perspective," *OECD Report on Quantum Computing*, 2020.
8. "IBM's Approach to Quantum-Safe Cryptography: Preparing for a Post-Quantum World," *IBM Research Whitepaper*, 2021.

10.2 AI-Powered Attacks and Defenses
References Footnotes

1. Ian Goodfellow and Yoshua Bengio, "AI and the Future of Cybersecurity: Threats and Opportunities," *Journal of Machine Learning Research*, 2020.
2. "The Rise of Deepfake Cybercrime: A Case Study," *Journal of Digital Forensics*, 2019.
3. "AI and Phishing: How Machine Learning Is Making Attacks Smarter," *Symantec Threat Intelligence Report*, 2021.
4. "Machine Learning in Cybersecurity: Detecting Emerging Threats," *IBM Security Whitepaper*, 2020.
5. "The Role of AI in Incident Response Automation," *McAfee Cybersecurity Insights*, 2021.
6. Miles Brundage and Shahar Avin, "Autonomous Cyberattacks: Ethical Challenges and Risks," *Journal of Cybersecurity Ethics*, 2018.

10.3 Securing Emerging Technologies: 5G, Edge Computing, and Beyond
References Footnotes

1. "Cybersecurity Challenges in the 5G Era," *European Union Agency for Cybersecurity (ENISA)*, 2021.
2. Nazli Choucri and David D. Clark, "Securing 5G Networks: The Geopolitical Dimensions," *Journal of Global Security Studies*, 2020.
3. Weisong Shi and Schahram Dustdar, "Securing the Edge: Challenges in Distributed Cybersecurity," *Journal of Cloud Computing*, 2019.
4. "The Importance of Zero-Trust Security in Edge Computing," *Forbes Technology Council*, 2020.
5. Rolf Weber, "The Internet of Things: Cybersecurity Challenges and Opportunities," *Journal of Information Security and Privacy*, 2018.
6. "AI and Machine Learning in Cybersecurity: Trends and Predictions," *Gartner Cybersecurity Insights*, 2021.
7. Don Tapscott, "Blockchain and the Future of Cybersecurity," *Journal of Blockchain Applications*, 2020.

10.5 Predictive Cybersecurity and Proactive Threat Hunting
References Footnotes

1. "CrowdStrike's Report on Predictive Cybersecurity," *CrowdStrike Threat Intelligence*, 2020.
2. "Gartner's Predictions for Cybersecurity in 2023," *Gartner Cybersecurity Insights*, 2022.
3. "The Role of AI in Predictive Cybersecurity," *Symantec Security Report*, 2021.
4. "Proactive Threat Hunting in the Modern Enterprise," *FireEye Mandiant Security Review*, 2021.
5. "The Importance of Threat Intelligence in Cybersecurity," *Mandiant Threat Report*, 2020.

CHAPTER 11
BUILDING CYBER RESILIENCE

11.1 Cyber Resilience
References Footnotes

1. "Cyber Resilience: Building a Secure Future," *European Union Agency for Cybersecurity (ENISA)*, 2021.
2. Paul Cichonski and Joshua Franklin, "Cyber Resilience: A New Approach to Cybersecurity," *Journal of Information Security*, 2019.
3. "Cybersecurity and Cyber Resilience: Two Sides of the Same Coin," *Gartner Cybersecurity Insights*, 2020.
4. "Measuring Cyber Resilience Using the NIST Framework," *NIST Cybersecurity Framework*, 2021.
5. "Cyber Resilience Metrics: Measuring the Ability to Withstand Attacks," *IBM Security Whitepaper*, 2020.
6. "The Role of Awareness and Preparedness in Cyber Resilience," *Symantec Cybersecurity Report*, 2021.
7. "Recovery as a Key Indicator of Cyber Resilience," *Verizon Data Breach Investigations Report*, 2020.
8. "Continuous Improvement in Cyber Resilience," *McAfee Cybersecurity Trends Report*, 2021.
9. "The Importance of Leadership in Building Cyber Resilience," *Accenture Cyber Resilience Report*, 2020.
10. "Challenges of Cyber Resilience in the Face of Advanced Threats," *Kaspersky Lab Threat Report*, 2021.
11. "Third-Party Risk Management as Part of Cyber Resilience," *Deloitte Cyber Risk Report*, 2021.
12. "The Future of Cyber Resilience in a Post-Pandemic World," *Gartner Cybersecurity Insights*, 2021.

11.2 Integrating Cybersecurity into Organizational DNA
References Footnotes

1. "Embedding Cybersecurity into Organizational Culture," *PwC Cybersecurity Report*, 2021.
2. "Cybersecurity as a Strategic Priority: Integrating Security into Organizational DNA," *Gartner Cybersecurity Insights*, 2020.
3. "The Role of Leadership in Cybersecurity Integration," *McKinsey Cybersecurity Report*, 2021.
4. "The Importance of Board Engagement in Cybersecurity," *Deloitte Cyber Risk Report*, 2020.
5. "Building a Security-Conscious Culture to Prevent Cyber Incidents," *Verizon Data Breach Investigations Report*, 2021.
6. "Security by Design: Best Practices for Developing Secure Systems," *NIST Cybersecurity Framework*, 2021.
7. "Aligning Cybersecurity with Business Objectives for Competitive Advantage," *Accenture Cyber Resilience Report*, 2020.
8. "Overcoming the Challenges of Integrating Cybersecurity," *Symantec Cybersecurity Trends Report*, 2021.

11.3 Adaptive Security Architecture
References Footnotes

1. "The Role of Adaptive Security in Cyber Defense," *Gartner Cybersecurity Insights*, 2021.
2. "The Evolution of Security Architectures: Moving Toward Adaptive Security," *Symantec Cybersecurity Report*, 2020.
3. "Adaptive Security in the Age of AI-Driven Threats," *Cisco Cybersecurity Trends Report*, 2021.
4. "Predictive Analytics and the Future of Cybersecurity," *CrowdStrike Threat Intelligence*, 2020.
5. "Adaptive Security: Combining Prevention and Detection for Better Protection," *Palo Alto Networks Security Whitepaper*, 2021.
6. "Real-Time Detection and Response in Adaptive Security," *Forrester Cybersecurity Review*, 2020.
7. "The Role of AI in Incident Response," *IBM Security Whitepaper*, 2021.
8. "Machine Learning and AI in Adaptive Security Architectures," *McAfee Security Insights*, 2020.
9. "Challenges in Implementing Adaptive Security Architectures," *Gartner Cybersecurity Insights*, 2021.
10. "The Future of Adaptive Security in Cyber Defense," *Gartner Cybersecurity Review*, 2021.

11.4 Continuous Monitoring and Improvement
References Footnotes

1. "Continuous Monitoring and Risk Management in Cybersecurity," *NIST Cybersecurity Framework*, 2020.
2. "The Role of Continuous Monitoring in Advanced Threat Detection," *Forrester Cybersecurity Insights*, 2021.
3. "Real-Time Analytics for Cyber Threat Detection," *Gartner Security Review*, 2021.
4. "Automated Response and Continuous Monitoring: Improving Cyber Resilience," *IBM Security Whitepaper*, 2020.
5. "The Importance of Continuous Risk Assessment," *NIST Cybersecurity Guidelines*, 2021.
6. "Continuous Improvement in Cybersecurity: Learning from Incidents," *McAfee Threat Intelligence Report*, 2020.
7. "Cybersecurity Maturity Models: A Roadmap for Improvement," *Deloitte Cybersecurity Review*, 2020.
8. "Overcoming Alert Fatigue in Continuous Monitoring Systems," *Symantec Security Report*, 2021.
9. "The Benefits of Continuous Monitoring for Cyber Resilience," *ENISA Cybersecurity Report*, 2021.
10. "AI and Machine Learning in Continuous Monitoring Systems," *Gartner Security Insights*, 2021.

11.5 Collaborative Defense and Information Sharing
References Footnotes

1. "The Role of Collaborative Defense in Cybersecurity," *ENISA Cybersecurity Report*, 2021.
2. "Collaborative Defense: The Power of Sharing Threat Intelligence," *Symantec Security Insights*, 2020.
3. "Improving Incident Response Through Information Sharing," *Palo Alto Networks Security Whitepaper*, 2020.
4. "The Importance of Sharing Threat Intelligence to Combat APTs," *FireEye Threat Intelligence Report*, 2021.
5. "(ISC)² Report on Cybersecurity Information Sharing and ISACs," *(ISC)² Global Cybersecurity Study*, 2020.
6. "The Cybersecurity Information Sharing Act (CISA): Facilitating Public-Private Collaboration," *DHS Cybersecurity Insights*, 2020.
7. "The Role of Public-Private Partnerships in Cybersecurity," *Deloitte Cybersecurity Review*, 2021.
8. "Building Trust in Cybersecurity Information Sharing Initiatives," *McAfee Threat Report*, 2021.
9. "AI and Automation in Collaborative Cyber Defense," *Gartner Cybersecurity Insights*, 2021.

www.ingramcontent.com/pod-product-compliance
Lightning Source LLC
LaVergne TN
LVHW061955050326
832904LV00008B/287